MANAGING SOCIAL SERVICE SYSTEMS

MANAGING SOCIAL SERVICE SYSTEMS

JOHN W. SUTHERLAND Rutgers University

PBI
a petrocelli
book
new york / princeton

Printed in the United States of America

1 2 3 4 5 6 7 8 9 10

Library of Congress Cataloging in Publication Data

Sutherland, John W
  Managing social service systems.

  "A Petrocelli book."
  Includes index.
  1. Social work administration. 2. System
analysis. I. Title.
HV41.S88   658'.91'361   77–21806
ISBN 0–89433–004–7

WITH GREAT APPRECIATION,
this book is dedicated to the memory of Professor Bill Baker
(late director of the Canadian Center for Community Studies) . . .
and to the continuing good health, good humor and productivity
of my most valued mentor, Professor Eric Trist

CONTENTS

PREFACE

A monarchy is a merchantman which sails well, but will sometimes strike on a rock, and go to the bottom; a republic is a raft which will never sink, but then your feet are always in the water.

Fisher Ames, 1795

Let's look for a moment at what Fisher Ames had to say so early in this nation's history. The implication is a comforting and fascinating one. The monarchy and its autocratic variants chart a course, adhere to it with efficiency and singlemindedness; but momentum may destroy them utterly. For the path of any nation is fraught with obstacles and perils. Run up against them at speed, and the vessel of state is lost. Thus, the republic—slow, clumsy, comical in its caution and conciliation—is the safer vessel. Being the victim of constantly competing currents and slowed thereby, it cannot run aground or strike an obstacle at any speed. It therefore remains afloat, even though its occupants may be always somewhat frustrated by its sloth and capriciousness of course. And they are always uncomfortable, always a little damp and drawn about the gills.

The raft of our republic is waddling and bobbing in crosscurrents of considerable strength these days. On the one hand, there are those whose course is set for an unlimited expansion of social services. As an ancillary interest, they are the advocates of civil rights, the enemies of racial and sexual discrimination, the opponents of capital punishment, the advocates of amnesty for draft deserters. Lending moment to the other current are those who are convinced that any expenditure on social services is sure to weaken the structure of the republic and imperil its buoyancy. They, by and large, are more concerned with property rights than with civil rights; are the opponents of affirmative action and racial quota systems; are opposed to forced busing to achieve racial balance; are for state's rights and limited central government; are for societal discipline and are proponents of responsible individualism. As is consistent with a reasonably mature republic, there are many individuals indifferent to the prophetic extremes. But among those at the opposite ends of the socioeconomic spectrum, there appears to be little room for reason or reconciliation. After all, ideological positions are not, as a rule, answerable to either logic or fact. But we need not be concerned with any grand societal classification schemes in this volume, merely with the social service sector as an axis of opinion and recrimination.

Again, our citizens are apt to see social service programs either as the highest achievement of national conscience (the most compelling evidence of a society's compassion), or as evidences of "softening of the brain" rather than "softening of the heart" (the vanguard of an insidious and sapping socialism). There are merits to both positions. It is certainly true that a nation which neglects the poor, the sick, the aged and the impaired has little to recommend it to history; but it is also true that when the social service sector absorbs too large a proportion of national resources—when it brings institutionalized paternalism to too large a proportion of the population—the integrity of that society is in danger.

For the scientist concerned with the social service sector the guiding issue becomes the concept of the tradeoff: how much to be allocated to the social service sector, how much to be retained for the uses of the mainstream society? But he will probably have to stop as soon as he gets started. What determines the tradeoff is not the kind of objective analysis for which he is equipped, but the natural forces of human interest and sympathy that largely elude our paltry analytical skills. Again, we are passengers on the raft. Were we aboard the sleek and tightly controlled merchantman of monarchy, autocracy or oligarchy, then the scientist could presume to fix a figure . . . so much goes to the social casualty, so much remains for the amusement or employment of the more advantaged. Although our econometric capabilities are hardly up to the task of fixing an optimal point of tradeoff, this probably would not stop the technocrat (at least it has not stopped those economists and social planners who direct the allocation of resources in centralized states). But being aboard this republican raft of ours, the scientist's task is simpler. He must suggest this: it is equally to the interest of both the advocate and opponent of the social service sector that it be managed in such a way as to produce maximal effect for the resources invested. More simply, whatever the absolute level of resources society determines should flow to the social service sector, the guiding commandment to the social service administrator is this: be efficient in the use of those resources, and be prepared to be accountable for your proficiency.

Efficient use of resources siphoned off to the social service sector will mollify the opponents, or at least take some of the sting out of the income we preempt. Efficient use of resources will also directly serve the benefits of those to whom social service programs are directed—the client. When we husband the resources properly and allocate them with discipline, the level of service we can provide the social casualty increases accordingly. In short, the incompetent social service administrator is as much the enemy of the social casualty as of the taxpayer.

Now, so far as I have been able to ascertain after a decade of observation, the quality of management in the social service sector is not what it should be (and my observations appear to be born out by the hit parade of administrative horrors to which the newspapers and waggish columnists are constantly adding). Sadly indeed, many of those currently directing social service programs—or occupying managerial positions at any level—appear to have either only the most moderate exposure to managerial technology, or none at all. Moreover, their appreciation of general economics is sorely limited and leads to constant miscomprehension of the implications of their positions and programs. Now some might complain that the managers in the private sector are not all paragons of proficiency either. There is merit to their point, but not for us here. For here we are not concerned with the production of shoes or computers or toothpaste. Rather, we are concerned with organizations where errors of judgment or technique hurt people, not merely profits. For those who join me in believing that people are possibly more important than profits, the requirements for competency among social service managers must be more demanding than those pertaining to their commercial or industrial counterparts.

I shall not belabor these points here (but I shall unabashedly raise them again, and more pointedly, in the text itself). Rather, I shall suggest briefly why it is this book was written and set out its ambitions. First, the motivation for this volume lies with the fact that the man or woman operating in the social service sector —or studying to do so—does not really have a wide range of literature to choose from. Indeed, the books that pretend to teach public administration are often reconstituted business administration texts, or variations on the civics books that we had in high school. This is in part understandable, for it is really only recently that public administration was realized as something *more* than business administration and something *different* than political science. So the social service administrator may, with great legitimacy, blame the academic community for having largely ignored his needs, and left him vulnerable and unsupported. Now, in a modest way, I hope this volume will help provide some technical support to the practicing manager and also to those studying to assume positions in the public interest. But these pages profess no panaceas; there are really no technical tricks that can substitute for the judgment, sensitivity, assiduity and integrity of the manager himself. He is the master, and the tools and procedures offered here must always remain his servants.

As for the arguments and discussion developed in these pages, I have tried to keep them as accessible as possible to those who have little technical or mathematical background. This volume aims to be blatantly pragmatic and of immediate service. Now as to whether this will actually be the case or not, I cannot answer. All I can answer for was my intention to be of use and my ambition to be intelligible to the widest possible audience.

In order to take our inquiry out of the realm of the abstract, I have made constant and thorough use of a real-world model. This model reflects a health care delivery system developed for Multnomah County, Oregon. It's intention was, first of all, to provide a working system for the delivery of essential medical services in Multnomah County itself. But there were at least two ancillary missions: (a) to construct a prototype for a national health care program, one which would go beyond the conceptual barriers that constrain much current thinking on that subject (e.g., where virtually all schemes proposed by congressmen and other interested officials are tied to an insurance-based or prepaid modality, which as we shall see is hardly a promising perspective); and (b) to show how a single set of managerial techniques could be used to implement an *integrated* social services program, one where welfare, medical care and virtually all other social services could be contained and administered within a single organizational framework. The hope is that this concentration on a working system—the illustration of how an actual delivery system was developed—will be more useful to the reader than would a long dissertation that dealt only with abstract principles or with the emergent but still immature "theory" of social services.

In summary, then, the fundamental path of inquiry in these pages is the way in which certain well-developed, essentially simple management science and system analysis techniques may be fruitfully, and immediately, applied to the social service sector. The intention is to be as practical as possible here, and this may often mean that certain technical equivocations and procedural qualifications will be dealt with most casually. Yet for those who are interested in the origin of the system design techniques that are introduced—or in the theoretical underpinnings of the various management techniques discussed—there are rather frequent citations and references given in the text itself.

Finally, in my constant criticism of the way things are now managed—in the sweeping indictments of social service functionaries—there is deliberate dramatic excess. Not all officials, by any means, suffer from the defects that are made so much of in this text. Many social service functionaries are individuals of great wit and character, and of enormous skill and intelligence. But this book is not only for them. It is also for those who know that there are gaps in their background and lacunae in the continuum of their capabilities. And the reader should really not bridle at the criticisms offered, for we are all of us babes when it comes to the comprehension and control of the social service sector we have evolved. I well know that success in social service management —like success in love and war—is predicated on the tiniest turns of fate and judgment, none of which can be completely programmed. But when we make the effort to pursue expertise, when we challenge our limitations and demand technical sophistication of ourselves, it is in the best possible cause. For compassion and benevolent intentions alone cannot soothe the lot of the social casualties looking to us for sustenance and support.

John W. Sutherland
Piscataway, N.J.

ACKNOWLEDGMENTS

This volume is the product of experience (not all of it successful) with about half a dozen different social service design projects. The most important of these, which I have drawn on in some detail in these pages, is the design of a new generation of health care delivery system for Multnomah County, Oregon (under HEW Contract RX-74-14). I am grateful to Dr. Hugh Tilson for his support and guidance in the development of this system design, and to Ed Ross of HEW for his interest and courtesy. To Jim Baker and Paul Molnar go special thanks; they were key members of the system design team, and Jim Baker is as close as we come these days to a true genius in system design. Finally, two colleagues who have also become great friends reviewed the original manuscript and served as constant sources of guidance and correction. Chris Nielson is director of a major health care delivery system and brought to these pages the benefit of his field experience and realism. Stephen E. Seadler, president of New York City's UNICONSULT Corporation, provided much advice of technical substance and, as is his way, tended to show me those many places where my enthusiasm outstripped my discipline. To all these fine men, I give my most sincere thanks. And then there is my university, Rutgers, the State University of New Jersey. No institution could more faithfully or substantially support the work of its members.

PROLOGUE

The social service administrator has a tough balancing act. On the one hand, taxpayers demand that he seek efficiency and economy of operation, that he run our health, education and welfare programs like a business. But when he attempts to act like a businessman, he often finds that efficiency can be purchased only at the expense of true effectiveness. And if the constant conflict between the ends of efficiency and effectiveness is not enough, he has a third focus of concern . . . client dignity. Moreover, unlike his counterparts hidden away in giant commercial corporations or huge government bureaucracies, the social service administrator is visible, susceptible to political manipulation and always vulnerable to embarrassment and personal recrimination.

When all these factors collide and interact, the managerial challenges facing the social service administrator are genuinely staggering. He is often driven to ask, in all seriousness, whether the social service systems we have evolved are really *manageable*.

The answer that emerges from these pages is this: social service systems are indeed manageable, but only when the social service administrator is able—and willing—to take full advantage of the potential offered by the modern management sciences. Which management science techniques are most useful—and how and when they should be employed—are the major foci of this volume. And the sum of all the arguments and discussions is simply this: the proper social service manager is one who tempers compassion with competence, and uses discipline and precision as the constant complements to intuition and imagination.

MANAGING SOCIAL SERVICE SYSTEMS

1

THE SOCIAL SERVICE CONTEXT

INTRODUCTION / The social service sector is distinguished not only by the nature of its mission, but by the prevailing characteristics and orientations of those staffing social service programs . . . or studying to do so. By taking a somewhat extreme (and perhaps unjustifiably critical) perspective, we shall find that a majority of social service functionaries are simply not technically qualified for the demands their positions exert. That is, they lack the necessary training in management science, have little comprehension of the economic constraints on their operations, and generally also tend to be confused about the relationship between the social service sector and the other segments of a society. In many cases, the neglect of the technological aspect of their positions stems from their belief that a good heart or benevolent intentions are sufficient for social service management. Not so! When we examine the properties of the social service sector, we shall find that administrative competence implies a much greater level of analytical sophistication than would be required of someone operating at a similar level in private industry.

To a great extent, the indifference to technological predicates —and the general neglect of economic criteria—are the major sources of the ineffectiveness and inefficiencies that abound in the social service sector. It must be suggested that the managerial problems will not disappear simply by trying to make social service management look more like business administration. There are critical differences, and the attempt to out-of-hand impose commercial criteria on the public sector is unlikely to be successful. Yet there are ample opportunities for the manager or administrator to amplify his skills, wit, capability and contribution. This initial chapter asks that the practitioner or student of social service management alert himself to the technological potential which may not have been a part of his college curriculum. And once alerted, the subsequent chapters attempt to lend substance and some immediacy to a technology of social service management.

THE SEVERAL SOCIAL SERVICE MODALITIES

Presumably, most readers of this book will already be predisposed to the idea that social services are a valuable adjunct to any nation . . . a societal imperative. In terms of the mechanics involved, the social service programs a nation supports are to be viewed in terms of three types of basic missions. First, some are programs of *containment* which are meant to control contrasocial or disruptive tendencies or individuals, such as institutions for the criminal, the insane, the retarded, the diseased. A second set of programs may be viewed as *maintenance* functions, as exercises in institutionalized charity; these provide certain necessities of life to those unable to acquire them on their own volition (e.g., welfare, food stamps, public housing, medical aid to the indigent). Finally, there are what may be called *developmental* social service programs. These, fundamentally, exist to amplify the capabilities of individuals to contribute to the economic mainstream, and thus serve as the basis for a social "investment." The Job Corps, certain retraining programs, compensatory education, preventative medical care, nutritional programs for children, etc., have a distinctly developmental predication.

The rationale for programs of each type is essentially different. Initially, containment programs are designed to remove unpleasant or dangerous influences from the mainstream communities. Prisons, mental institutions, sanitoriums, and reform schools, for example, are viewed by the public as proper public institutions, and are deemed to be of some explicit value to the community at large, irrespective of any philosophical considerations. Containment programs, on the other hand, are seen (at least by the economically initiated) to be the collective price a nation pays for the continuation of a capitalist economic system. For capitalist systems, with their emphasis on competition and productive fitness, are bound to leave some proportion of the population unemployable and therefore penurious. But because no nation any longer advocates social

Darwinism* (where noncompetitiveness is seen to be a natural phenomenon not to be interfered with or ameliorated), some portion of the resources generated by the productive sector of the nation are transferred via maintenance programs to the unproductive. That such programs exist may or may not be testimony to the conscience of the productive individuals in a nation. In some cases, it is merely testimony to the general affluence of a system, such that some portion of the collective income may be diverted to maintenance functions without putting any severe strain on the amenities available to productive members of society. When, however, the amount diverted to maintenance programs becomes significant, their continuation will in large measure indeed depend on the sympathy of the resource sector or at least on their humanistic impulses.

Finally, no nation has as yet paid much attention to developmental social service programs. In the first place, they generally require some method of organization other than strict bureaucracy; particularly, they demand considerable personal, face-to-face interchange between professional and client. Secondly, we simply have not evolved—through the offices of the social and behavioral sciences—an effective set of developmental strategies. That is, we really have no reliable technology for resocialization, socioeconomic energization, politicization or behavioral (cognitive) maturation. Thirdly, developmental programs imply significantly high transaction costs; that is, they require rather significant expenditures of time and resources per individual. Another reason may be this: we simply have not been able to adequately articulate the economics of social development. There are, of course, certain vague assertions about the fact that a well-fed, well-educated individual makes a better worker and citizen, and is therefore not likely to demand any continuing support from maintenance programs . . . that a little bit invested now can save enormous amounts later on. But given the status of modern industry (with its high capital/labor ratios and consequently its decreasing demand for labor relative to total population), these economic arguments do not carry much influence. Moreover, and possibly this is the critical reason why develop-

*This does not, however, mean that there are not strong pockets of individuals whose ideological referents owe their substance to Social Darwinism; it may still be considered the major referent among the more affluent and active members of both American and European society.

mental programs have received so little attention, most members of the productive population really think that welfare and other maintenance programs are developmental. They simply do not realize that the welfare system, as it is currently constituted, is not really intended to "develop" individuals. Rather, it simply serves to institutionalize poverty and dependence. In short, maintenance social service programs have the net effect of creating an artificial, secondary economy which is more or less hidden —but no less permanent and entrenched—than the productive economy to which a majority of citizens belong.

In general, then, arguments supporting the social service sector will tend to be either economic (materialistic, quasi-rational) or axiological (based on emotional or affective predications). The economic arguments simply suggest that containment programs serve the cause of societal stability and security; that maintenance programs are necessary because all men are not created equal in terms of their ability to contribute to the mainstream economic system; that developmental programs are useful because they are an investment in human resources which will eventually prove to be a source of net savings to society, by dampening the need for containment or maintenance programs. The axiological arguments are more popular.[1] They usually articulate a moral or humanistic basis for social service programs. They evoke the sympathy (but seldom the empathy) of the productive members of society, reminding them of their Judeo-Christian duty to those less fortunate. The axiological arguments are thus variations on the Sermon on the Mount, and are often components of what has recently come to be called the liberal persuasion. But for reasons we shall shortly discuss, the typical liberal tends to become a vocal advocate of the maintenance aspect of social service programs, a critic of the containment functions, and generally indifferent to developmental emphases. The net result, then, is that maintenance social service programs, and the population that has to be "maintained," always increase.

It is fair to suggest that we are now at a watershed in many of the modern industrial nations. The flow of resources into maintenance programs has reached the point where the economic viability of the primary economic system is imperiled by capital shortage and erosive tax rates. In addition, the expansion of these programs may be seen to constantly increase the number of government dependents in a nation, which quickly translates into an erosion of individual initiative and dignity and into a captive source of political support for liberal politicians. These considerations are in part responsible for the increasing threat of a middle-class backlash, as seen recently in several European countries and as forecast for the United States. Such a backlash could promise the therapeutic effect of forcing our attention away from maintenance programs and toward developmental social services. But it is in the nature of reactive social movements that they tend to be indiscriminate and too enthusiastic. The danger is that the positive aspects of social service programs will be discarded along with the distinctly dysfunctional components.

For in the matter of social service programs, there is little moderation. Those functionaries whose sympathy or sustenance depends on social service programs argue for their expansion, and condemn all those demanding moderation as illiberal or inhuman. The conservatives demand, on the other hand, a return to some sort of unmediated "rugged individualism." The positive point of tradeoff—the rational appreciation of the social service sector—must rest somewhere between these extremes. This is essentially the point that President Carter tried to make when, in his inaugural address, he asked for social compassion to be tempered by governmental competence. And this is mainly the point we shall be looking at in these pages. For when social service programs preempt too many individuals and too many resources, they become dysfunctional and a source of socioeconomic enervation. But to revert to the catch-as-catch-can, solipsistic society advocated by strict conservatives is not an attractive or particularly sophisticated alternative.

The strategy which this volume will recommend, then, is twofold. First, it will ask that those who advocate the unlimited extension of maintenance programs rethink their position and try to moderate their predispositions. It will ask the same thing of those for whom all social service programs are anathema, and who are preconvinced that we are not our brother's keeper in any sense at all. Second, there is the matter of tempering compassion with competence. This basically means two things: (a) the student of social services and those of a priori liberal persuasions are going to have to educate themselves to the economic implications of social service programs; and (b) those whose ambition is to manage social service programs are going to have to recognize that it takes more than a good heart and benevolent intentions to administrate them successfully. Simply, the social service manager is going to have to educate himself to some of the technical potentials of modern management science. To the extent that social service functionaries continue to be largely ignorant of economics and indifferent to management technology, social service programs will continue to be sources of irritation and embarrassment. The incompetent social service manager wastes resources that could otherwise be available to ease the plight of those for whom he pretends such affection, and thus contributes directly to the probability that social service programs will be curtailed rather than increased.

To begin, then, we must understand something about the peculiar sociological and economic concepts that affect a majority of those staffing—or studying to do so—the social service sector. For the generally immoderate social philosophy held by many social service functionaries, coupled with their general neglect of economic precepts, translates directly into factors which tend to make the social service sector either inefficient or, more seriously, a direct threat to the integrity of the nation supporting them.

A SOCIOLOGY OF THE SOCIAL SERVICE FUNCTIONARY

Generalizations are dangerous, but sometimes useful. They are dangerous because they ignore the exceptions in the cause of sweeping categorization. But categorization is one of the primary instruments of the social sciences. The categorization developed here—a straw man, as it were—will be a deliberately critical and pessimistic portrait of the social service functionary. All the criticisms will not apply to any single individual, and indeed there may be no individuals at all who share all the adverse attributes being set out. But these adverse attributes must be identified and forwarded. The reason is this: as already suggested, the analytical demands and managerial challenges facing the social service administrator are staggering. Given the complexity of the milieu in which he must operate, the administrator may be forgiven some errors of judgment and analysis for complexity implies error! But there are certain perspectives and prejudices held by some social service functionaries which, a priori, cause them to neglect the potential for rationalized actions and optimal decisions. These prejudices and perspectives, by and large, become intelligible to us as components of the *sociology* under which many social service managers and students labor, often tacitly. And this sociological referent that is so widespread in the social service sector is not attractive; it tends to consistently elevate compassion beyond competence and to spawn managers who are more attuned to affective than scientific management (and, correlatively, more reactive than reflective). These attitudes and imputed prejudices do not remain with the individual. Rather, they become institutionalized and impose themselves on social service programs, forcing before-the-fact anomalies and inefficiencies. Thus, the properties of the social service sector are no less unique—no less definable—than the properties of the individuals it attracts. And like most of us, operatives in the social service sector have both assets and liabilities. So, the attempt here will be to give an admittedly artificial hypercritical portrait of the type of individual who seems to dominate the population of social service students.

Initially, those entering public administration programs—in their social services emphases—will tend to have undergraduate degrees in one of the social sciences. This means that, for the most part, their intellectual preparation is concentrated in subjects such as sociology, political science, social psychology, even anthropology. They will, as a group, have had only moderate exposure to any formal economics (though they may have had the usual macro-micro course). They will, moreover, have had only a passing familiarity with statistical processes, their exposure being restricted mainly to operations with simple inference and sampling, with perhaps a token glance at correlation and regression. But, as a rule, they will not have had any working exposure to time series analysis (algebraic or spectral, nonlinear), to Bayesian statistics or to the theory of experimental design. As a group, they cannot work well with factor analysis, cluster analysis or analysis of variance. Very few will have had any exposure to mathematics or formal logic, and may indeed have a dread fear of such rigorous subjects.

Now, coming from such limited introductions to the social sciences, their appreciation of scientific methods will tend to be limited largely to rhetorical constructs or to strict empiricism. The rhetorical component emerges because so much social science is conducted by individuals perhaps more interested in changing the world than in understanding it. It is beyond the scope of our inquiry here to conduct a detailed criticism of rhetorical social science, yet an enormous literature exists elsewhere for those to whom it is of interest. The empiricist basis of their social science work will reflect the general interest in hypothesis testing . . . experimenting with generalizations of limited scope and substance. But the rhetorical orientation of much social science makes itself felt even here, for the experiments that are designed and implemented tend most often to be directed at a limited "proof" of an assertion that may be conditioned by the investigator's own predilections.

It is not surprising, then, that most graduate students in social service curricula think of themselves as individuals with a *mission*. This mission, when generalized, takes the form of a more or less active liberalism.[2] In short, many students tend to see themselves preparing for a career in social prophecy or reform. This is conditioned by the fact that most students see the government as the agent of social change, and see the private sector as reactionary or obstructionist. Their general perspective is that the primary economy—staffed by corporate or commercial interests and their acolytes—is the cause of severe *asymmetry* in the distribution of economic (material) prerogatives across the population as a whole. Their general picture of the economy is that a few families and corporate entities preempt wealth and income at the expense of a significant proportion of people falling into certain obvious minority categories (e.g., blacks, Chicanos, the aged, women). Without fully comprehending the implications—and usually without any strong theoretical command of the economics involved—the student offers himself as an advocate for the various forms of *transfer* of wealth and income. In short, the majority of those preparing for degrees in social service administration seem to see the most interesting and the most essential function of government as the redistribution of wealth and income. For the social service sector, when institutionalized, is the mechanism by which this redistribution takes place. True to their primarily rhetorical backgrounds, the value or morality of redistribution is not at issue—it is automatically accepted and defended as an intuitively good strategy.

Their enthusiasm for their chosen careers is, however, somewhat dampened by a couple of realizations. Initially, they are aware that the majority of those holding posts of significant influence in government have a rather different preparation than that which they are obtaining. Particularly, senior functionaries come primarily from business or legal backgrounds. There-

fore, the social service student expects to play a subordinate role
. . . as an implement rather than an implementer or formulator
of policy. Their expectations are also dampened by a knowledge
of the close interconnections between business and political interests. Therefore, most students tend to focus their attention on
local opportunities and seem to be generally preoccupied with
community or urban systems. The net result is that students in
social service curricula do not seem to have the signal ambition
and drive that so often characterizes students in graduate schools
of business or in law schools. Moreover, students in social service
programs seem to come from families that would tend to fall into
the lower-middle or upper-low classes; thus, their social sophistication and their optimism about careers seems to be perceptibly
lower than that of students in business or legal programs. The
average social service student is a bit more humble, a bit more
complacent and perhaps a bit less personally secure or presump-
tuous than his counterpart in the technical or professional
schools. And as is to be expected, the general proportion of
women and minority students is higher in most public administration programs than in most engineering, law or business programs.

In somewhat kinder and more positive terms, the average
public administration student tends to think of himself as involved, sympathetic, liberal, humanistic and real. But we must
caution that this latter attribute does not mean *realistic.* Rather,
being real in the present context means behaving according to
certain social conventions: pursuing affective relationships; being
willing to recite one's limitations; seeking emotional gratification
within the context of some organized movement. In short, they
are quite susceptible to institutional settings and collective ambitions. The result is a strongly idealistic component in their character, a kind of temperate utopianism. But here there is a striking
anomaly: they are also fascinated by the highly personalistic,
influence-based systems of local and state government. That is,
they tend to see governmental functions and offices as extensions
of individual personalities and are constantly referring to the
realities of political life. And what this interest in realities means,
when translated into operational terms, is this: they are predisposed to compromisive or suboptimal solutions and are usually
prepared to subordinate their idealism to *practical* considerations. In other words, even as students, they display a certain
innocent cynicism and are not convinced of the feasibility of

optimal or rational criteria. Thus, their idealistic and cynical references are constantly in conflict. We often find them trying to ingratiate themselves with the very political machinery which, more often than not, retards the reforms which their idealism applauds.

Their idealism—coupled with their general lack of any technical background—also serves to make them the enemy of technocratic or patriarchal institutions. They fear and distrust the FBI, the CIA, the military and even the police . . . and even the hard sciences. They tend to see all such agencies as extensions of corporate or material interests, as the vehicles for discrimination and counterreformation and exploitation, or as artifices that defend the positions of societal elites and which challenge social evolution.

Indeed, it may be suggested that the majority of students in social service programs tend to have a distinctly *matrist* point of view, which contrasts strongly with the stern patriarchal orientation of capitalist-Protestant society.[3] In particular, they tend to view the individual as a victim of circumstances and chance, and therefore neither to be punished for his transgressions or failures nor applauded for his successes. The basic presumption is that society at large is substantially to blame for social casualties, the criminal, the poor, the ignorant, etc. The derivative strategy, then, is this: one may improve human nature by improving the environment. And improving the environment means implementing a wide range of social service programs (medical care, compensatory education, rehabilitation, welfare, etc.). In short, the student sees himself and the social service system as a surrogate parent, doing for the disadvantaged what good families would do for less fortunate individuals. The important point here —and the one which causes such conflict between social service functionaries and other sectors of society—is this: the social service functionary always looks for an exogenous (outside) cause for any social casualty. For example, the criminal is a criminal because his father beat him or because his mother was a drunk; the black man is unemployed because he is discriminated against; those on welfare are there because labor unions are exclusive and try to restrict membership; the indigent elderly are in that sad state because corporations tried to maintain cheap labor; the inner-city child cannot read or write because property taxes tend to discriminate unfairly against schools in poor neighborhoods; the unemployed are jobless because industry tries to discipline the labor sector by inaugurating periodic recessions, etc.

The members of the employed middle class, the legion of blue-collar workers and conservatives in general contradict this position by suggesting that people are poor because they are lazy or stupid, that criminals are criminals because it's easier than working, etc. The truth, of course, lies somewhere in between. It is no more useful to assume that all individuals are victims of outside forces than to assume that environment plays no role at all. But we have to understand that the average social service student has a very real stake in the concept of nonautonomous man and in egalitarian movements in general. For, again, he is not likely to have any kind of inheritance; he is not likely to see himself as ever becoming wealthy; he is not preparing in the skills required by a technocratic society; he probably does not have a network of useful social contacts in the higher levels of either private or public enterprise; and he is not likely to have much faith in the ability of any single individual to rise above his circumstances or to author his own conditions. Rather, his ambitions are generally quite humble, and as such he is one of those most likely to benefit from any egalitarian trends (and he does not picture himself as likely to be impaired by any compensatory or antidiscrimination legislation, such as the progressive tax system, the affirmative action program, etc.). Thus, his own prospects tend in large measure to condition his sympathy for reform and his animosity toward traditional values. It is not merely moral or philosophical considerations, then, that find him opposing capital punishment and supporting the Equal Rights Amendment, that find him arguing against tax loopholes and for the negative income tax, that find him an adversary of defense expenditures and an advocate for food stamps.

This portrait of the typical public administration student will probably change in the next few years. There are basically three reasons for this. First, the number of jobs available in the public sector (relative to those available in business or the standard professions) is increasing. This means that some individuals who would be highly qualified to assume other professions are, by default, going to enter public service, and some of these will serve in social service programs. Second, the changes in socioeconomic ideology afoot since the 1960s have tended to equip many young people with an empathetic interest, whereas earlier generations were pushed toward an ego-orientation as the moral attribute for capitalistic, individualistic society. Third, the disparity of income opportunities between governmental and commercial sectors is gradually reducing, such that public service no longer carries the automatic material penalties it once did. There is soon going to be an increase in the number of technically qualified, highly alert and tough-minded young people applying for admission to public administration programs. And there is yet another reason which is even more interesting: many individuals —and in many cases these are the very best students available— realize that the managerial aspects of the public sector provide the highest intellectual challenges. Thus, curiosity and determination is leading some highly qualified students into the public sector and away from engineering and business administration.

But as things now stand, the majority of social service students subscribe to a most damaging and naive proposition: that the faults of society are due solely to faults of character. The implication is this: all that is required to improve society is to get rid of the crooks and the selfish businessmen and public officials and replace them with men of good will and benevolent disposition. They look at industry and see anticompetitive practices, profit mongering, environmental despoliation, discriminatory hiring, etc., and see the solution in terms of a new morality. They see a self-absorbed, generally uninvolved public—indifferent to the plight of the poor, the noncompetitive, the minorities—and they see salvation in the wakening of a social consciousness. In this charming but innocent set of assumptions, the student of social services is joined by a majority of his faculty instructors. For, like their students, rhetorical social science faculties will tend to have little if any background in quantitative methods or formal epistemology.* They too will tend to feel ill at ease when reminded of the technical predicates of modern management. Their discomfort will often translate itself into a very definite defensive behavior, where they will seize upon vague opinions about the inadequacy of technical instruments or quantitative management schemes and precondemn—while being essentially ignorant of the implications—attempts to bring formal analytical structure to public administration. The result is a rather pathetic reinforcement of the students' inadequacies by an equally impaired faculty. And given this sweeping neglect of administrative science and formal analytical procedures, the issues of social ser-

*As most of my social science colleagues will agree, rhetorically oriented social scientists still tend to predominate, even though some social science has become highly quantitative indeed (though not always to the benefit of the subject).

vice management—along with issues in public administration in general—continue to be condemned to the realm of rhetoric.

This abject neglect of the technical requirements of management in the public sector is not malicious. It is simply that the social service student and the usual professor of public administration are caught at the watershed. The capacities of quantitative management are exploding with significance and implication, and this significance and implication escapes those who were educated before administrative science had emerged as a relatively mature movement. Those professors whose own educations were essentially rhetorical—rooted in normative political science qua civics or in the amorphorous "principles of management" which were popular well into the 1950s—simply pass their deficiencies onto their students. Not to do so, to demand of their students a capability which they themselves do not possess and cannot judge, would be impractical.

The net result of all this is that social service curricula tend to be shy on technological substance, and thus social service programs continue to be ill managed. One thing is absolutely clear: as much societal waste, deprivation and dysfunction is caused by faults of mind—limitations of intellect—as by faults of character. Despite the self-serving litany that social service students and professors offer, a good heart is no substitution for analytical sophistication or intellectual discipline. To those who suffer from poorly conceived, ill-executed social service programs, it makes little difference whether those responsible are real villains or merely well-intentioned incompetents. It hurts just as much either way.

So the sermon offered here (and which this volume will, it is hoped, bring to substance) is this: the challenge for the social service student is to continue to be sensitive and sympathetic, but to become technically competent and analytically disciplined as well; to be empathetic, but also facile with the instruments of modern management science. If they fail to do this, the waste and inefficiency of social service programs will sorely dilute their effectiveness, and exacerbate the situation of the very social casualties toward which the student pretends such concern. But this assertion cannot really be defended until we have explored the economics of the public sector. For it is the economic realities which set the demands on those aiming to manage the social service sector, and these same realities which, if unrecognized, become the most potent enemies of meaningful social evolution.

THE ECONOMICS OF THE PUBLIC SECTOR

The rather peculiar "sociology" of social service staff and students leads to an equally peculiar interpretation of the role of economics in the public sector. It appears that many individuals feel that economic criteria are either irrelevant to social service operations, or inimical. Those who find economic issues irrelevant suggest, in the main, that the central purpose of economics —the rational allocation of scarce resources—simply does not apply in the public sector. Two reasons are offered for this remarkable position. First, it is suggested that rationality is unachievable in public or social service enterprise because of the inability to measure outputs or to control specific costs. Again, this is a position taken by the technically naive, for certain cost-benefit and other surrogate operations are indeed available to discipline both cost and output formulations (some of which will be dealt with in later sections). But to those who believe that rationality is unachievable, the optimality promised by strict economic procedures is out-of-hand denied, and social service administration becomes all art and no science. A second argument advanced by those who suggest that economics is irrelevant to public operations is this: the criterion of scarce resources does not apply because the public sector can "create" its own resources. In short, we don't need to worry about husbanding public resources, because the government can expand the resource supply at will. It is perhaps this view which most confounds businessmen, economists and fiscal conservatives when dealing with social service functionaries and with many congressmen and federal officials as well. Indeed, it is a sad indictment of the economic literacy of the public sector that so few officials have any strong background in economics to complement their careers as lawyers, druggists, soldiers, ministers or what have you.[4]

The more interesting position with respect to economic criteria is that taken especially by those involved with social service programs. While many functionaries in other sectors of public enterprise ascribe to the concept that economics is irrelevant (for reasons just cited), many social service functionaries feel that economics is *inimical* to their mission. That is, they actively resent and repress any intrusion of economic criteria. This is only to be expected, considering arguments about their beliefs and background developed in the previous section. For the social service functionary feels that it is a crime to try to put a price on human dignity, on human welfare, health or contentment. Thus, when the economist or the businessman suggests that there are limits to the ability of the working population to support an indigent population—that the costs of social service programs should be controlled or curtailed—those in the social service sector condemn him as insensitive or illiberal. Again, the lack of much formal exposure to economics, the compensatory mission of social service programs and the inherent distrust of private enterprise that affects most social service functionaries leads them to be the enemy of any attempt to force a tradeoff between the need to care for social casualties and the need to conserve economic resources. In short, it is distinctly in the interest of social service functionaries to relegate economic criteria to a distant second place. This neglect of economic criteria leads to social service programs which are consistently suboptimal (and to conservatives, abjectly profligate).

Now, were economic criteria really irrelevant or really inimical to social service missions, then we would have no complaint about their neglect. Nor, really, could we criticize the widespread practice of appointing people to social service positions who have as little exposure to economics as they do to formal management science. But economic criteria are the *central facts* of social service programs, even if they are transparent to most social service functionaries. There are definite limits to the resources which may be expended in the public sector, and there is therefore a desperate need for adequate controls and rational allocation mechanisms. But because these economic criteria are difficult to comprehend and not immediately tangible or visible, their impact is little understood. Perhaps, however, Figure 1.1 will help.

Initially, it is necessary to understand that any nation has a finite set of *real assets.* In very abbreviated terms, these all reflect the ability of the nation to produce a valuable product. Real assets are thus associated with the primary (agricultural, extraction), secondary (manufacturing, assembly, processing) and tertiary (service) industries the society supports. A unit of currency represents a share in this productive capacity. In actual practice, the real value of a unit of currency, relative to the value of other nations' currency units or with respect to some measurement standard such as gold, represents the ability of that unit to be traded for real goods or services. And when considering productive potential rather than real assets, the value of a unit of cur-

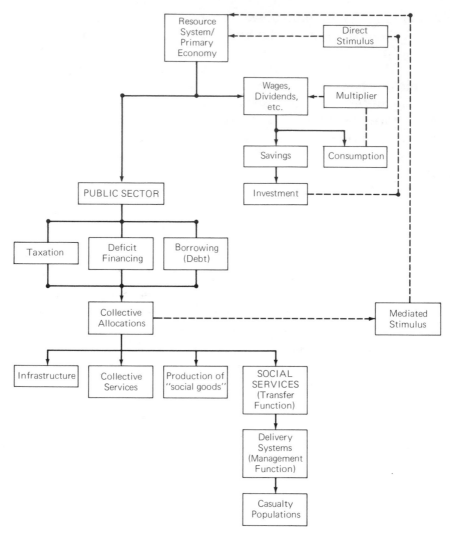

FIGURE 1.1 / The resource "map"

rency represents the *expectations* about the health and integrity of a nation's productive base. Now, when we couple this real asset base with the characteristics of the human population (particularly its ability to do productive work, or its willingness), we get what is referred to in Figure 1.1 as the *resource system*. A crude measure of the value of the resource system is Real GNP (the value of the goods and services produced during some interval, after adjustment for inflation). In general, the GNP is distributed to the factors of production (labor, management and capital owners) in the form of wages, benefits, interest, dividends, etc.

Any recipient of any income increment may elect to save it, invest it or consume it. This is as true of corporations as it is of individual employees. Now, as for savings, there are basically two modalities. First, there is the type of savings that takes resources away from the resource base. Particularly, these are funds that are stored in the mattress, sent overseas for concealment, etc. Such savings reduce the money supply and hence the ability of the economic system to generate capital. Institutional savings— money put into savings accounts or into bonds, etc.—are made available to expand the resource base. Particularly, such savings are available to support mortgages, to provide equity or debt financing for new industrial projects or to provide consumer credit for the purchase of automobiles, furniture, boats, etc. Corporate entities or other types of commercial firms have the equivalent of a savings function in their retained earnings accounts . . . funds that are pumped back into the enterprise to provide financing for future operations or to expand operations. It is important to understand, then, that productive savings become the source of *investment* stimulus to the resource system and provide for the expansion of the economic potential of a nation.

Now, again, in addition to the two types of saving functions, there is the option to *consume* one's income, or to exchange it (or some asset) for goods or services. Thus, expenditures on direct consumption serve to *clear the market,* to provide a payment for goods or services that, for the most part, are already produced or which will be produced on demand. Thus, expenditures on consumption do not, of themselves, serve to expand the economic system or the resource base. But they perform a critical function nevertheless. Every dollar directed toward the purchase of some good or service may be viewed as a vote for that particular product, and may thus serve to improve the probability that its supply will be expanded. But it should be clear that to actually expand this supply—to create incremental goods or supply incremental services—investment capital is required. And as was seen, that investment capital is generated by the savings function. It goes without saying that as investment increases, the number of jobs available to a nation's citizens increases, as does the GNP. Therefore, expansion of the economy is a reinforcing function, either upward or downward. As the savings ratio increases and as business draws upon these resources to expand production, the demand for employees increases, which in turn increases the level of consumption, etc. But the spiral works negatively as well; to the extent that the capital is not available to support industrial expansion, or to the extent that the "price" of capital (the interest rate) rises, the expansionary stimulus is dampened and unemployment results, etc.

Continuing with the explanation of Figure 1.1, we now have to pose the public sector as being in competition with the resource system (the private sector) for capital. Particularly, in addition to being able to spend or save a dollar, the individual and the corporation or productive agency must also be prepared to pay a certain increment of earnings or wealth to the government. In short, the public sector has a claim on a certain proportion of GNP, depending on the structure of the tax rates and tax provisions. As the figure indicates, taxation is only one of the mechanisms by which the public sector captures resources. The other two are debt financing (public borrowing) and deficit financing (an artificial increase in the money supply).

The *debt financing mechanism* is the collective counterpart of the business investment function. The government or any fiscal authority may borrow from the public to finance the expansion or delivery of public goods and services. The public buys treasury bills, E-bonds, etc., and this money then becomes available to the general treasury and may be used to support any of the several functions listed under collective allocations in Figure 1.1. The point is this: funds which flow to the government through the debt financing mechanisms are funds not available for business expansion, at least not directly. But we shall explore this point more thoroughly in a moment.

Now the *deficit financing mechanism* has no real counterpart in the private sector (though certain forms of securities manipulation yield essentially the same effect). Particularly, it is not to be confused with the mechanism of consumer credit (where the consumer is allowed the use of funds which are already generated and which presumably have some correspondence with the real asset base). Rather, deficit financing simply involves an increase in the money supply, without a corresponding increase in the asset base. Its effect is insidious and generally misunderstood. What it does is this: it reduces the value of all other units of currency, because a "share" in the real asset base represented by a dollar is now eroded by the currency creation process. Specifically, we now have more dollars but no increase in the value of goods and services available. Therefore, deficit financing always leads to inflation and cannot do otherwise.* And inflation merely represents a *hidden tax* on the population of the nation at large. It simply means that every dollar of income, every dollar of savings, is now worth less than it was before. In order to compensate for the reduction in the value of the unit of currency, the producers of goods and services raise their prices so as not to erode their own income or profit status. Thus, the same number of dollars buys fewer goods or services. The effect is just the same, then, as increasing the tax burden on the individual . . . he has less real income that he may dispose of as he sees fit. And the critical concept that the student of social service administration should be aware of is this: whereas direct taxation is a progressive mechanism—burdening the well-off relatively more than the poor—inflation as a symptom of deficit financing is a distinctly regressive mechanism. It hurts the poor relatively more than the wealthy, because the highest price increases will generally tend to appear in those commodities with the least elasticity of demand—the necessities of life.† Food, clothing, shelter and energy become the first targets of inflation; the poor spend a much higher percentage of their income (whether from earnings or welfare, etc.) on these necessary commodities than do the more wealthy individuals. Therefore, when we accede to the financing of public programs or governmental functions through the deficit financing mechanism, we are doing a great and immediate disservice to those toward whom social service functionaries profess such concern and sympathy.

The reader may find this assertion a bit confusing, for he is constantly exposed to a contrary viewpoint: that deficit financing of government programs actually stimulates an economy. This logic is very deceptive, however, for reasons already suggested. Deficit financing artificially inflates the money supply, which in turn deflates the value of each outstanding unit of currency. Suppose, as is a currently very popular ploy, that we decide to

---

*At least in the short run. The presumption is that as the economic cycle turns up, the public sector (via taxation) will run a surplus, which is then used to reduce the deficit. But as will be shown, the very fact of a prior deficit acts to make a subsequent surplus improbable, and the Keynesean logic here falters badly.

†Inelastic commodities are those whose demand schedule does not vary significantly with changes in the price level. They are, as a rule, those commodities for which there are no effective substitutes, or which are deemed more or less essential to a certain standard of living (thus, even some "luxuries" may be inelastic in these terms).

expend the funds secured through deficit financing by providing public employment (or, as an ancillary program, increase allocations to the unemployed or to those on welfare or social security, etc.). Now it would appear that this influx of funds into the hands of people who will spend them would increase the income available to the nation's consumers, and therefore increase demand for goods and services. This should cause business to expand their productive capacities by hiring new workers, etc. In turn, this should reduce the population of unemployed and reduce the welfare rolls by offering new job opportunities in the private sector. This, essentially, is the logic behind what public officials call *pump priming*. But in reality, this dynamic does not materialize.

Rather, businessmen recognize the increase in demand associated with deficit spending for what it is . . . artificial and impermanent. Moreover, the inflationary response kicks in very quickly, raising prices to compensate for the reduced value of the currency. Therefore, they may meet the incremental demand by drawing down on inventories and not creating any new product. When this occurs, it is now possible to have high inflation (fueled by the artificial increase of the money supply) accompanied by substantial unemployment (which reflects the pessimistic interpretation of the businessman about the permanency or substantiality of the demand increase). Until recently, such a combination was thought to be impossible. Indeed, it had long been suggested that unemployment and inflation would have to move in opposite directions, such that an increase in inflation would engineer a decrease in unemployment, etc. But in this traditional analysis (summarized in what is known as the Phillip's Curve), the effects of deficit public financing were not really considered. So a new term has had to be entered in the economist's vocabulary to express the state where unemployment and inflation are both significant—*stagflation.*

Now, while deficit financing is not the only villain in this picture (the intractability of labor unions and profiteering among business corporations and speculators fuel the fire), it provides a sufficient cause of stagflation. Therefore, an *increase in public expenditures does not act as a stimulus to the private economy when the increase is financed from deficit sources.* If the government official or congressman cannot distinguish between one dollar and another, the private investor and corporate executive can. But, when financed through the mechanisms of taxation or debt financing (borrowing against real assets), then public expenditures do have an expansionary effect. However, the critical question here is this: Is the effect of mediated stimuli (public expenditures mediated as it were by collective allocation decisions) the same as the direct stimulus affected by private savings and individual investment decisions? The answer is simple: it cannot be! The reason is the one given earlier that the effect of dollars saved and dollars spent is different. And the government is a consumer of resources, not a saver or producer. Therefore, when the government taxes or borrows from the private sector, these funds are thus expended to clear the market for collective goods and services, the public counterpart of the individual purchases made by private consumers. Such expenditures may be considered in the same light as private consumption decisions: they tend to increase the probability that the good or service purchased will be expanded. But, again, this cannot be done unless the investment capital from private savings is available. To the extent that governmental tax rates or borrowing demands are high, this detracts from the available investment capital base and therefore acts to make real expansion improbable.

To make this point somewhat clearer, look at Figure 1.1 again and note the various classes of collective allocations made by the public sector. Initially, central authorities may choose to invest in infrastructure of various sorts (roads, rails, ports, dams, etc.). Infrastructure generally supports the mainstream economy through the advancement of transportation, communication or other basic functions. The general rule is that collective allocations for infrastructure investments are made when a broadly distributed need exists, but when elements of the private sector are not able (or motivated) to provide the investment. Infrastructure investments generally provide benefits on three dimensions: (a) they put people to work, and therefore expand the demand for goods and services from the mainstream economy (by increasing national income); (b) they provide the prerequisites for more efficient industry and commerce, and thereby serve to stimulate general economic development within the private sector; (c) and, indirectly, they enhance the quality of life for society as a whole, to the extent that transportation, communication and other facilities are expanded. Thus, collective investments in infrastructure do have a stimulating effect on the general economy, at least up to a point.

In the category of collective services are basically the informational, managerial and security functions of government and its satellites. In short, collective services house the business of government and become comprehensible as investments in planning, coordination, control, etc. When evaluating benefits of investment in collective services, two types of returns are considered: (a) the direct stimulus which is provided by the earning power of government employees, which translates into demand for goods and services from the private sector, and some portion of which may flow into savings to stimulate capital formation; (b) the benefits from the "managed" environment which government provides its citizens and economic sector, which in turn is said to provide the controlled framework within which the private sector may operate.

Next there is the set of *social goods* which the public sector may produce. Here, the public authority acts as a substitute for private enterprise or as its competitor. For example, when a railroad or a steel mill is nationalized, the government now is in the business of producing a product which would be capable of being produced by the primary economy. It is very common to find quasi-nationalized operations in all countries, such as nonautonomous utilities (the TVA, for example), agencies like AMTRACK, subsidized and controlled airlines and shipping concerns, etc. The economic integrity of such enterprises must always be adjudged with respect to their efficiency as producers or users of capital, much as we would seek to evaluate the marginal contribution of any private concern. And just like private enterprises, they generally provide benefits on three dimensions: (a) they provide employment, and therefore stimulate effective demand for other goods and services; (b) to the extent that the social goods produced are interrelated with other functions, they exert a leverage on ancillary production by other concerns (e.g., the AMTRAK operation or certain nationalized utilities really provide a direct stimulus to industry, much as would pure infrastructure investments); (c) to the extent that the product produced is an inherently valuable one, investments in social goods add directly to the real asset base of the society as a whole.

Now, in contrast to these three types of public enterprise (infrastructure, collective services and production of social goods) is the *social service sector*. This, basically, exists to implement a transfer of resources from one sector of the society to another. Specifically, it forces a redistribution of income (and, to a lesser extent, wealth) from the better-off members of society to the worse-off. And as already suggested, the social service sector as it is currently constituted is primarily a maintenance mechanism. Particularly, a considerable majority of the collective allocations that flow into the social service sector are used for the provision of basic subsistence or consumptive functions for the indigent; that is, the funds go to support basic medical care, nutrition and shelter, leaving the social service client with a very small (and occasionally no) discretionary income. Therefore, the benefits associated with social service programs become intelligible in terms of just a single effect: an increase in the demand for products and services associated with the transfer payments awarded to social casualties, and the demand stimulus of the salaries paid to those who staff the social service sector. The stimulus provided by social service transfer payments largely ends at this point, for in almost no case will recipients of social service support translate any of their income into savings or other investment forms. Thus, the social service sector, from the standpoint of strict economics, represents the least attractive destination for a nation's resources, *at least when the maintenance aspect of social service programs is stressed to the virtual exclusion of real developmental thrust.*[5]

Now, the reader has no doubt been deluged in recent years with dissertations proclaiming the value, necessity and moral imperative of social programs. Indeed, anyone who presumes to criticize welfare, unemployment, social security, food stamps or any of the other programs of this genre is often automatically condemned as inhumane, reactionary, exploitative or insensitive. But, let's take a quick look at some of the arguments *against* social service mechanisms and against forced income transfer in general:

1. The existence of a threshold criteria which provides minimum sustenance and support for indigents artificially reduces the labor force available to staff the more menial and less attractive jobs in industry and commerce (and in public enterprise as well). To attract employees to menial occupations, industry is required to inflate the salary rate above that which would prevail in the absence of transfer payments. Particularly, these jobs must offer salaries which exceed (after taxes) the level of support that one may obtain by not working at all. These salaries are artificially high because, it is argued, the net productive effect of menial jobs

(the profit impact) is less than the threshold wage. There-fore, the existence of the welfare base means a certain level of misallocation of resources by the private sector . . . over-compensation at the lowest levels of enterprise.

2. The transfer mechanism itself is condemned for its dampen-ing of the capital formation process necessary for real indus-trial expansion and therefore necessary for real economic growth. Particularly, transfer payments tend to take discre-tionary income away from those who would have saved a portion of their income (and thereby made it available for investment) and give it to those who will generate no sav-ings, to those who must consume all their income. There-fore, the social service mechanisms tend to act as a *capital sink*, where the expansionary potential of the mainstream economy is drained away, thus diluting its ability to provide jobs and generate profits and pay taxes. And as already sug-gested, this capital and investment-reduction effect will tend to become even more severe when social service pro-grams are financed from deficits.

3. Maintenance-oriented social service programs may retard capital formation in another way. Particularly, as it is ar-gued, the existence of a welfare "cushion" and of unemploy-ment and Social Security benefits, etc., has the tendency to make individuals more profligate with their salaries. That is, it leads them to consume a greater amount of their income than would be the case were the cushions not available. The individual no longer feels it is necessary to put aside a sav-ings base against layoffs, the proverbial "rainy day," and against old age or retirement. Therefore, his consumption function increases and his propensity to save decreases. When this occurs on a large enough scale, the nation's aggre-gate savings ratio decreases, and the basis for investment capital formation is eroded. (This situation has not really emerged in much force yet, but as the social service um-brella expands to cover the individual against more catastro-phies—e.g., medical conditions, structural unemployment—it is suggested that the reduction of savings will eventually decline quite substantially.)

4. It may also be argued that the social service sector exacer-bates general inflation by concentrating demand for the

basic necessities in an *artificial* market. For example, welfare clients generally tend to be housed poorly in facilities that charge the clients more than could possibly be gotten from an independent tenant. The same thing is true of the grocery stores, liquor stores and other retail operations which cluster around social service or welfare clients (e.g., grocery stores in ghetto areas charge more for basic food products than do stores in the better neighborhoods, and their higher insurance premiums account for only a relatively small proportion of the incremental difference between ghetto and mainstream stores; some medical practitioners have distinctly exploitative operations set up in areas having a high proportion of public dependents). It is suggested that the very existence of social service programs tends to breed an inferior and insidious commercial satellite, charging prices and delivering a quality of service that simply would not be tolerated by better informed, more mobile consumers.

The realities of the social service sector may then be very different than the claims which are often made for it. The social service system may not, indeed, serve to ultimately increase the labor force, but may actually contract it; the payments which are made to social casualties may not in truth act to amplify real demand, but may often be preempted by distinctly unproductive or even illegal markets. Finally, because the entire income of most indigents is used to support day-to-day consumption, there is simply no direct economic stimulus associated with this income because there is no savings (and, moreover, unprincipled ghetto businesses and other agencies which prey on the disadvantaged simply tend to adjust their prices upward in the face of any increase in the level of social service support; that is, there is a localized inflation which assures that, despite increases in transfer payments, the public dependent receives no real increase in goods and services). Thus, as the social service sector expands in a neighborhood, there is not even any pressure exerted to expand local retailing operations. In short, the net economic effect of social service programs is apparently distinctly negative. Still others argue that the net social effect of these programs is also negative, tending to create a passive, unmotivated and self-regenerative subculture that saps the character and ambition of those susceptible to it.

This, then, is the sorry portrait of the social service sector as the relatively objective economist might paint it. The criticism can take another direction as well. Again with respect to Figure 1.1, note that funds that are allocated to support social service programs pass through some sort of *delivery system* before they get to the client or social casualty. At the federal level, the delivery system may collectively be called the Department of Health, Education and Welfare, and the agencies of the other governmental departments which effect transfer payments (e.g., the food stamp program of the Department of Agriculture; the Veteran's Administration). Agencies or bureaucracies with similar functions exist for every state and, to a lesser extent, for every county and municipality. When considering the delivery systems, we are always interested in what is called the "transaction costs" of social services: *the proportion of the transfer payments* (the total budget for social service programs) *which do not reach the client.* The transfer costs are resources that are used up in the administration of the programs, the funds which are used to support the agencies themselves. For some agencies and programs, less than 30 percent of the funds allocated for social services actually find their way to the client or social casualty (a figure which, incidentally, is even worse for some private charities). To the extent that waste and inefficiencies exist in delivery systems, the already somewhat deleterious effects of social service programs on the general economy is exacerbated, and the effectiveness of the programs from the client's standpoint is much reduced.

Thus, we can summarize where we are. First, social service programs represent a moral or axiological commitment by a society to those who, for some reason or another, are unable to participate in the mainstream socioeconomic system. Social service programs cannot generally be defended on economic grounds unless they are of the developmental genre; maintenance programs, almost without exception, can be shown to be a net detriment to the economic sector and therefore a misallocation of societal resources. As is entirely within the prerogative of any society, however, it may elect to subordinate economic issues to those of conscience or emotion or whatever. But in doing so, it should recognize that nothing comes for free: that for the gratification of our social conscience, we must be prepared to pay an economic price. Therefore, the purpose of this volume becomes explicit: it is to investigate ways in which a society may meet what it defines as its social obligations, but at the same time minimize the disadvantageous economic effects. Now, basically, there are only three ways this can be done:

1. First, and certainly most propitious, is to shift away from the emphasis on containment and maintenance programs, and to thereby move from a welfare to a *development* posture. In this case, transfer payments going to support social service programs would not simply be viewed as resources going down the capital drain, but as resources being devoted to an *investment in human resources* that will have distinct returns for the primary economic sector ... healthier, better educated, better adjusted and better motivated individuals to staff both private and public enterprise, and to thus enhance the real asset base of the society. But as things stand at the moment, most agencies which refer to themselves as "human resource systems" are really just brokers for collective capital and have no viable developmental impact.

2. A second strategy would be to distinguish between those social casualties that are remediable and employable, and those that are chronically unsuited to any type of employment (the insane, the aged, the disabled, etc.). The result would be to put all employables to work on public projects, such that society at large gains some direct benefit from its transfer payments (e.g., expansion of infrastructure), and welfare supports for this group becomes transformed into proper salaries (and with developmental programs in force, the public employment program could become a most positive mechanism). In this way, a subset of the indigent population contributes to the real asset base of the society, rather than attenuating it.

3. The shift to a developmental emphasis in social service programs—and the institution of the demand for some return for public support from the population of employable indigents—are policy issues whose resolution awaits us in the future. But there is an alternative available which will immediately make the social service sector less a detriment to the general economy. This is to increase the efficiency of delivery systems and to reduce the transaction costs associated with those maintenance-oriented programs which

society is determined to support. To the extent that we can affect real savings on the administrative dimension, the economic "drag" of the social service sector on the mainstream economy is decreased, and the material integrity of the nation as a whole is thus increased. Or, what is perhaps even more important to many readers, administrative efficiencies may translate directly into an increase in the quality of service that social programs can provide their clients.

Now, as was suggested earlier, it is this third focus—social service administration—which concerns us in this volume. Before beginning to show how adequately efficient programs might be designed and implemented, it is worthwhile spending a few moments on the matter of management integrity in the social service sector.

MANAGEMENT INTEGRITY IN THE SOCIAL SERVICE SECTOR

It was Coventry Patmore who once suggested: "Nations often die of softening of the brain, which for a long time passes as softening of the heart." The essential message for us here is that even those functions performed as evidences of humanism or sympathy must be tempered by a great deal of intelligence. Simply, the most essential criticism against social service programs is that, when carried too far—when they consume too great a portion of a nation's resources—the fundamental viability of that nation itself is endangered. This, of course, is the sad lesson that England has learnt, and essentially the lesson that, in the most direct way possible, is now being taught to New York City.

This brings us back to a central theme: it is not enough to place men of goodwill and benevolent intention in offices of administrative responsibility; they must also be men of managerial sophistication and analytical aptitude. It is thus that the concept of managerial integrity, as it applies to the social service sector, has a dual focus. On the one hand, the manager should be a man of good character, free from the venality, egoism, arrogance and politicism that has apparently affected so many who have gone to work in the public sector. But he should also be a skilled administrative technician, wise to the implications of economics and well schooled in the promises and potential of management science. Again, to paraphrase a thought of President Carter, public officials must temper their compassion with a healthy dose of competence.

This brings us directly to the matter of the kind of intellectual preparation the social service administrator requires. And it brings us squarely against a troublesome anomaly. The three major conduits by which individuals reach positions of public responsibility (political science, law and business administration) all suffer from certain defects. Particularly, neither the political scientist, the lawyer nor the businessman is entirely ready to meet the demands of management in the public sector. The first of these conduits—the public administration program based mainly on descriptive political science—is often devoid of both technological and normative substance. If one had to characterize their mission, and admittedly adopting an extreme position, such programs seem to exist primarily as a vehicle by which the student is introduced to the *realities* of the public sector. Of course, realities is itself a rhetorical term, which can mean many different things to different individuals. In the context of the public administration program that derives its substance from such political science, it generally means this: the exposure to the student of the way things work . . . to the structure and dynamics of existing or historical governmental systems. In such programs, as is really to be expected, the political considerations tend to receive more attention than the administrative implications. To some degree, such programs become an elaboration on high school *civics,* and they tend to be staffed more by individuals drawn from the ranks of retired (or reconstituted) political functionaries than by academicians, per se. The substance of such programs thus tends to be rooted strongly in reportage and anecdote with prescriptions made available largely as a distillation of personal experience. As a result, there is more concentration on the way things *are* done than on the way they *should* be done, and these programs thus tend to become largely "vocational" (even when they benefit from being housed within a prestigious university). To survive with this orientation, the program must have strong links to operating administrations at the local, state or federal level. These links provide the fringe financing for the programs and also provide a pathway to employment for graduates. And the critical quality of this linkage means that such programs often become the servants rather than critics of highly placed public functionaries. The net result is that graduates of such programs tend to have essentially the same limitations of technique—and the same behavioral referents—as do current practitioners; they thus serve as "substitutes" for existing public administrators, and not as an evolutionary force. So, for the graduate of the public administration program that is a spin-off from political science faculties, management often becomes all art and no science.

The second of the conduits which sends people into public service is the law. Indeed, at the highest levels of government—including both elective and appointive office—lawyers dominate with great consistency. But we must ask about the relevance of legal training to social service operations or to public administration in general. When one asks this question, he will probably receive an answer something like the following: "The lawyer is especially well qualified for public administration positions because of his ability to efficiently obtain a large amount of information from a large number of different sources and to generate recommended prescriptions in the form of briefs. Simply, the lawyer is a quick-study expert in evidence, and generally sensitive to the consistency of compromise required in the public sector. Moreover, he has learned to reason logically and is thus presumed to be immune to the ideological or invested biases that so often sway the common man toward irrational positions and erroneous actions." Now this portrait of the lawyer is perhaps too kind, for there are several attributes which act to make the average attorney something less than an unadulterated blessing when he accedes to a managerial role. First, we must understand that the study of the law often emphasizes *semantics* at the expense of *syntax*.[6] In great summary, this means that lawyers are often better at performing simple exegesis than at conducting substantive and formal analysis, per se. Second, the lawyer may have a tendency toward management by precedent and a well-developed (and not entirely unhealthy) distrust of innovation, experiment and heuristic problem-solving procedures. Third, as to the matter of his quick study and his tendency to answer issues with articulate briefs, there is a serious problem: most significant public issues simply cannot be resolved by expedient secondary research or by solicitation of expert opinion. In short, the lawyer is usually not conditioned to the patience and humility that are the primary constituents of all proper and productive science, or to the probabilistic rules of evidence that apply to socioeconomic as opposed to juridical phenomena. As a fourth defect, the lawyer's penchant for compromise—while nesting well with the political scientist's concept of democratic government as a conciliator of separate and inimical interests—may often be unsatisfactory in a managerial contact. In almost no case will a compromise solution be an effectively optimal one; compromise usually excludes the probability of being completely wrong, but often only at the expense of being completely right. Finally, the lawyer generally lacks any direct and significant exposure to management science, per se, and may also have only the most casual knowledge of the substantive bases required of proper public administrators: economics, sociology, behavioral science, history, etc. As a result, the lawyer offering his services to the public sector may often be long on wit, charm and connections, but short on information and technique.

The third of the conduits that send a stream of functionaries to the public sector is business administration, either as an activity or a curriculum. It has long been a presumption of management theory that a good manager is portable . . . that he can just as well manage a government bureau as a chicken ranch, or handle a welfare office or social service agency just as well as he can handle an oil refinery or a shoe factory. This somewhat arrogant presumption is now going out of fashion, yet the peripatetic executive—shuffling back and forth between commerce and government—is still very much with us. As a rule, the businessman —or the individual trained in business administration—will have at hand some concept of how to actually coordinate and control a complex enterprise. Those business administrators who have been trained in the newer and more sophisticated curricula (those that supplement the largely rhetorical principles of management* with statistical, system and mathematical subjects) are going to have a technical base that will prove of considerable value.

Indeed, we find an increasing number of schools of general management offering simultaneous preparation for positions in the public and private sector (and many major business schools, per se, have satellites that are dedicated to the public sector). This broadening of the management studies perspective is en-

---

*As a rule, rhetorical management curricula are distinguished by their insistence that management is the "art of getting things done through people," and, as a consequence, their heavy concentration in quasi-behavioral science. This is to be contrasted with the political science predilection for defining management as the "distillation of experience," and the consequent concentration on structural or contextual (e.g., realistic) subjects.

couraging, for it means that a population of technically initiated practitioners is being made available for public service. But the question is this: just how consonant is the technology of business administration with the demands of the public sector? The answer is disturbing, for public administration is something *more* than business administration. Therefore, the technological envelope within which the sophisticated business student exists must be expanded as he moves to the public sector. In short, public administration is not just another name for business administration, and the businessman offering his services to the public sector cannot legitimately pose himself as society's immediate savior . . . he will have a little learning to do. But he is welcome indeed! For his basic analytical and operational skills are set; true, they must be modified a bit, but he brings to the public sector an analytical sophistication that the political science student or lawyer will be hard-pressed to match.

But what of his technical deficiencies? What is this "more" that he must be prepared to do when he moves from the private to the public sector? The answer is brief, mainly because no one has really yet exhausted all the differences between the public and private sector (and also because many of the differences that we do know about are rather subtle and would demand much more space than is available here). First, the businessman loses one of his most powerful allies: the concept of profit. Profit gives the businessman a usually accessible, measurable and manipulatable index of performance against which to evaluate decisions and performance. Given an operative profit potential in an enterprise, the businessman may evaluate his alternative investment and allocation decisions according to their projected profit impact. Now, admittedly, this is not done with consistently great precision by businessmen; as commercial organizations increase in size, the ability to develop profit projection for major investment issues decreases rapidly. Nevertheless, after all is said and done, profit at least allows us to establish a clear-cut criterion for decision making, and allows us to measure the performance of one firm—or one manager—against that of another. Thus profit, like share of market, is a *reflective* criteria; it can tell us whether one organization or one decision maker is relatively better than another (given certain *ceterus paribus* assumptions), and thus becomes comprehensible as a unit of ordinal (ranking) significance. It cannot, of itself, tell us whether any particular firm or any specific decision maker behaved optimally during any interval. But in the competitive commercial world, optimal criteria are often ignored in favor of reflective criteria, though many scholars are unhappy about this.

Now, many of the most powerful and pragmatic instruments in the repertoire of the business administrator demand the existence of a profit base . . . marginal analysis, capital budgeting, etc. Many of the generic analytical instruments available to the modern manager imply a specific, measurable (dimensional) variable, such as the dollar. A majority of business analysis and decision tools also presume the existence of an output (product) that can be inventoried,* assigned a specific value (either the profit contribution via the revenue it will yield, or the real cost of producing it via cost accounting conventions). When the business manager moves from the commercial to the social service sector, he is largely deprived of profit and inventory as referents, and thus faces a potentially very much more confounding and ill-behaved context.

Now it may be argued that the concept of profit has a counterpart in the public sector, namely *surplus.* Surplus may be defined as the incremental increase in the value of an output relative to the inputs from which it was produced. When cast in dollar terms, surplus becomes profit. In the public sector, the implication is that surplus becomes intelligible as marginal *benefits.* [7] Thus the business firm computes profits (according to one accounting convention) as the difference between cost and revenue. It is then argued that surplus in the public sector becomes intelligible as the difference between costs of providing a service and the benefits that service yields to the population toward whom it is directed. It is thus that cost-benefit ratios have been proposed as the surrogates for the cost-revenue functions used by business firms. It is here that the lack of an inventory comes to haunt the public administrator. While revenue is a stable, precisely measurable quantity (which can be arrived at either arbitrarily by the firm as a pricing decision, or left to the free market mechanisms to compute), benefits are inordinately difficult to measure. What, for example, is the surplus produced by a school? Do we measure it as the number of students educated at a certain arbitrarily set level of proficiency? Do we assess it not in terms of numbers, but in terms of quality? Is, for example, a graduate

*Thus, commercial or private "service" industries share many of the problems of the public enterprise operating under a service posture.

engineer worth more than someone who graduated with a degree in medieval French poetry? Is a lawyer worth half as much as a chemist? Twice as much? Or do we try to measure educational benefits by tracing the careers of graduates over long intervals of time? If we elect to use distant measurement horizons, how can we use this information to make timely adjustments in our curricula? Essentially the same problems emerge when we try to establish benefit levels for hospitals or police departments, etc. Do we measure benefits in terms of the number of patients treated or in the number of criminal convictions obtained? Or, rather, should we measure benefits of medical systems by the lack of illness, and the performance of law enforcement agencies by their ability to prevent crimes? Are these omissive criteria accessible to us? If not, does not the measurement of benefits for such programs carry an enormous subjective or speculative component? If the subjective component is admittedly high, how can we legitimately make public administrators unequivocally accountable for their performance?

Instead of troubling the reader by reciting the other problems associated with cost-benefit or cost-effectiveness criteria, it is merely suggested that the business student's instruments and predilections do not fully prepare him to move into the public sector and immediately go to work injecting that technological component which is so needed. Rather, the businessman will find that the neat, deterministic algorithms and formulae he was taught in business school cannot be applied directly in the public sector. Rather, he will have to expend considerable energy and imagination in bringing the concept of "surplus" to operational significance, and enormous patience in trying to work around the lack of inventory bases. Thus, it is not at all true that the salvation for the public sector is simply to run it more like a business. The analytical challenges—and indeed the context itself—are in a very much different form in the public sector. Yet the businessman's basic instinct is good! The public sector—particularly the social service sector—can indeed benefit from the introduction of proper reflective measurements and from the imposition of cautiously applied surrogates for profit and inventory. It is merely that the vocabulary and procedures cannot be borrowed out-of-hand from the private sector. Moreover, the businessman is going to have to be prepared to make some behavioral adjustments were he to move to the public sector. Particularly, he is going to have to be prepared to be less reactive (as is useful in those engaged in competitive enterprise) and be more reflective. Secondly, he is going to have to be less concerned with how to gracefully recover from a mistake, and more concerned with how to avoid mistakes in the first place (for while errors in judgment in the private sector usually only hurt profits, mistakes in the public sector hurt people). Thirdly, because benefits or "outputs" are so difficult to measure in the public sector, he is going to have to be willing to be judged as much on propriety as on performance (in short, means are as important as ends in the public sector).[8]

To give a very brief idea of the *foci of managerial competence* in the social service sector, consider Table 1.1.

Although we shall have an opportunity to deal with these key ratios in more detail later on, we should have some tentative idea of their implication even at this beginning stage of our work. Initially, the determination of a particular (P/GNP) ratio is a political decision that sets the highest order constraint within which the social service sector must operate, reflecting the intrusive capability of government itself (in the tax rates it is able to enforce and the deficit financing load tolerable to the private sector). The second ratio is the direct constraint on the social service sector as a whole, representing the total flow of resources to that sector as opposed to the other possible destinations for public resources (infrastructure, collective services, etc.). The determinants of this particular ratio are complex: (a) the political pressure which the social service interest groups are able to exert on legislative authorities; (b) the state of the private economy with recessive tendencies developing a greater absolute need for transfer payments; (c) the degree of empathy inherent in the resource sector, determining the tolerable income-redistribution algorithm, etc.

But for any given social service program, it is the third ratio which is of critical importance, the resource-client ratio (X/Y). Indeed, it is the dynamic properties of this ratio which is the key focus for the performance of the program's executives . . . its senior management. Particularly, the central responsibility of senior social service management is to ensure a continuing increase in the resource-client ratio, such that the per capita service level of the program may be secularly increased. There are, however, some qualifications we should raise here. First, this is a "local" management objective, and not one consistent with proper economic theory. For the rational manager would not

TABLE 1.1 / The Structure of Managerial Responsibility

Foci	Ratio	Significance
The ratio of preempted to total resources for the t'th fiscal year	$\left[\dfrac{P}{GNP}\right]_t$	Measures the claims made on total resources through the taxation and deficit financing mechanisms, and usually reflects the prevailing politico-economic ideology of the nation.
The ratio of preempted resources devoted to the social service sector	$\left[\dfrac{S}{P}\right]_t$	Sets the transfer payment function and reflects the axiological predications of the society.
The ratio of resources to client population for the i'th social service program	$\left[\dfrac{X}{Y}\right]_i$	Sets the potential ceiling utility for any individual social service program by establishing the maximum per capita service level.
The ratio of transaction costs (administrative overhead) to line expenditures for the i'th social service program	$\left[\dfrac{T}{X}\right]_i$	Measures the administrative efficiency of the delivery system, such that $(X - T)$ becomes the set of resources effectively available to the client.
The cost-benefit ratio for the i'th social service program	$\left[\dfrac{X}{U}\right]_i$	Measures the actual utility delivered to the client population (in terms of impact) given the resources expended; the cost-benefit ratio is thus the surrogate for the "profit" function that disciplines units' performance in the private sector. This is the aggregate ratio, determined as: $\Sigma\ (X_j/U_j)$, where j is an individual unit.

raise the resource-client ratio to the point where diminishing marginal returns set in (see the second section of Chapter 2). Second, for the social service manager whose reference is the resource system (the taxpayers) rather than the casualty or client population, he may perceive his mission to be the gradual reduction of the resource-client ratio, and the consequent easing of the tax burden. Third, for some programs of a developmental nature, there is the explicit mission to gradually make clients self-sufficient, which implies a steady decrease in the resources allocated to any single individual.

In practical terms, however, these qualifications tend to be of negligible significance. In the first place, as regards diminishing returns to scale (or more broadly, optimal scale of plant or enterprise), such calculations can seldom be made with any authority or precision. It is a rule of socioeconomics that, for example, most commercial corporations tend to consistently exceed optimal scale of plant, and to this extent represent suboptimal organizations from the standpoint of strict economics. But in the social service sector, with the exception of the gigantic federal bureaucracies themselves, most programs are chronically underfunded (which means that diminishing returns to scale are not anywhere on the manager's immediate horizon). As for the efficiency-oriented social service manager, he will quickly find that it is basic structural and design faults which preempt resources in the form of system overhead, and that reductions here are generally more important than reductions in the funds that actually get transferred to clients or are actually transformed into services provided clients. Therefore, the manager who thinks he will best serve the taxpayers interests by reducing the resource-client ratio is generally mistaken; he would usually do better to clean up his administrative house first. Finally, developmental programs whose ambition is ultimately to put themselves out of business find that this development of human resources is enormously more expensive than maintenance. Thus, of the three strategies available with respect to the resource-client ratio listed below, (c) is almost always the most favorable:

(a) $\dfrac{d(X/Y)}{dt} < 1$ This implies that the per capita service level is being eroded, for the client population is increasing faster than resources, or resources are being reduced at a more rapid rate than the client population. In short, there is less service available for the average program client through time.

(b) $\dfrac{d(X/Y)}{dt} = 1$ The resource-client ratio remains essentially stable, indicating that no change in potential level of service is occurring (e.g., resources and client population are either increasing or declining proportionally, or remaining stationary).

(c) $\dfrac{d(X/Y)}{dt} > 1$ The resource-client ratio is improving, such that the potential level of service available to the average client is increasing through time.

The logic behind posing this ratio as the performance focus for senior program management is simply this: within an extremely large range, the effectiveness of any social service program is directly determined by the level of service the program can deliver to its client population. That is, until diminishing returns to scale set in, "more is better!" It thus becomes the responsibility of senior program management to see that the resource base grows faster than the client population or contracts more slowly. In practice, this means that program executives must strive to break the formula-budgeting (or categorical-funding) constraints under which most social service programs operate, for the formula-budgeting, categorical-funding mechanisms simply ensure a proportional rise in resources relative to the rise in client population. The central command for social service executives (policy makers, chief operating officers, agency director, etc.) is thus the following: *Maximize unallocated per capita resources.*[9] It is only by incrementally increasing unallocated resources (e.g., noncategorical resource support) that the resource-client ratio may be increased.* It is unallocated resources that provide educational institutions with the luxuries of more laboratory or athletic facilities per student or higher than minimally dictated teacher-faculty ratios; it is unallocated resources which allow a social welfare agency to reduce its case-load quotas and amplify individual attention; it is unallocated resources that permit hospitals to provide a larger equipment base per patient or to have a higher than minimally legislated employee-patient ratio. As the reader might note, this dictate really has no counterpart in the commercial or industrial world and therefore represents a managerial focus somewhat unique to the social service sector.

Moving on, one of the key determinants of the efficiency of any social service program is, of course, the ratio of administrative overhead (transaction costs) to total program resources, (T/X). As a general rule, efficient managerial systems for any given social service context will result in a smaller value for this ratio.

*Note that unallocated resources from one accounting convention becomes a special type of "overhead" that is available for expenditure at the discretion of the local management, and thus tends to increase the amenities associated with some service operation and may be used to increase level of service (in this regard, see the discussions in the first section of Chapter 5.)

It therefore becomes the key to the technical competence of the program management, whereas the (X/Y) ratio may merely reflect their skill as fund raisers, propagandists, lobbyists, etc. Therefore, one of the major technologically dependent functions of social service analysts is to find ways to *minimize the administrative overhead the management system exacts.* For the smaller this ratio, the greater the proportion of total allocations that can actually be used to generate a therapeutic impact on the client population.

If senior management (the chief operating executives) are for the most part responsible for the client-resource ratio and the transaction cost ratio, the operating administrators (the line managers) become directly responsible for the cost-benefit ratios their individual departments or units are able to achieve. Thus, the functional administrator in the social service sector (the line manager) becomes intelligible to us as the counterpart of the "cost-profit centers" which pertain in the commercial and industrial world. In other words, his local operation has certain *impacts* it should be making on the properties of the client population, and it has been given (or may command) a certain level of resources to generate those impacts. In the industrial sector, the manager is given a certain budget (inputs) to generate a certain output (revenue or value-added). His profit performance is the ratio of value-added or revenue to inputs. In the social service sector, the concept of revenue or value-added is generally intelligible in terms of *benefits,* which merely reflects the inability to index social service outputs with precise dollar values. But the principle is the same: the lower-level or line manager is accountable for the net effect of the resources allocated him. It is just that the concept of "profits" and the more fundamental concept of "surplus" are not usually accessible to us, or they are replaced by surrogates in ways which will become apparent in later chapters. Finally, as is obvious, the aggregate cost-benefit index for any social service program is the accumulation of the cost-benefit functions obtained by its subunits. Therefore, internal resource allocation decisions—the determination as to what units get what resource supports—should be disciplined by the apparent cost-benefit ratios for those units (just as allocations among social service programs as a whole should be dictated by the cost-benefit ratios the individual programs can promise). In this way, the social service sector becomes susceptible to the central commandment of managerial economics: *allocations of resources should be made in such a way as to consistently minimize opportunity costs.* In short, this means that every investment made in either a social service program as a whole or in any operating unit of any program should be such that there were no better way those resources could have been spent. It is this dictate which, of course, provides the technical substance behind the program budgeting and zero-base budgeting schemes which are now so popular in the public sector (and which have their original inspiration with the capital budgeting systems employed in the private sector). But in terms of analytical requirements and technological instrumentation, capital budgeting systems will look very different from the capital budgeting mechanism to which businessmen have been accustomed.

This is just a very brief overview of the technology of social service management, and thus sets one of the major foci for our concept of managerial integrity . . . the demands for competence which can legitimately be made on policy-level and operating managers in the social service sector. It makes most explicit the rationale behind the suggestion that social service management is as much science as art. Now, were this volume to be a treatise on management science or a purely theoretical effort, it could try to suggest why the instruments of public administration necessarily imply more statistical and mathematical sophistication than most common business administration tools can offer (why, for example, linear programming algorithms, most PERT formulations, most Markov constructs, most queuing models, etc., cannot be used as exclusive sources of decision discipline in the public sector; why, for example, capital budgeting techniques cannot really be employed without major modification in public enterprise; why traditional management information systems and even exotic cybernetic schemes have only a minor role to play in the normal business of social service administration; why, in addition, the envelope of certainty that higher-level decision makers impose on lower-level decision makers in private enterprise has no counterpart in the public sector; why the separation between policy and operational decisions, so popular in the corporate world, does not apply in the social service sector, where every decision automatically carries policy significance). But the ambitions for this volume are less august. The remaining pages will be used to show how certain well-defined, relatively widely distributed management science techniques may be used to design and implement a social service delivery system that has some probability of proving positive, if not optimal. Also shown will be how issues of dignity and conscience can be superimposed on economic criteria, and how proper tradeoffs can be developed between apparently competitive ends.

It is enough to close this introductory chapter with a brief reconsideration of managerial integrity. It can now be agreed that a good character is a necessary but not sufficient condition for managerial integrity. The sufficient condition is that the manager also be intellectually equipped for the tasks he has been asked to undertake, that he complement his sensitivity and humility with a strong technical capability. The point, then, is this: as far as integrity goes, neither the political scientist, the lawyer nor the businessman is able to meet *all* the demands that public administration makes without some additional preparation, some amplification of what he already knows. True, the man trained in the technology of commerce is likely to find this augmentation a bit easier and quicker. But the public sector needs the realism, contextual sensitivity and historical perspective that the political scientist can offer; yet if the political scientist is to accept a managerial assignment, he owes his constituency an effort to learn some management science. The public sector can indeed benefit from the precision and skepticism of the lawyer and from his determination to resolve conflicts with rationality rather than recrimination. Were he, however, to pose himself for an administrative post, integrity would demand that he search for a technical competence and probably also scan the modern social science with alertness and energy.

As suggested in these pages, then, the ideal social service administrator is going to be grounded in the social sciences, and particularly in political science; be responsive to the uses of procedure and conciliation that mark the lawyer's lot; and be comfortable with the kind of technical skills that are now demanded of graduates from the better business schools. Encouragingly, there are already some schools of public administration that demand these attributes of their graduates or that at least encourage their pursuit, but no one is demanding that the public administrator be all things to all men. Rather, all that is asked is that the man or woman seeking a position in the public interest cherish their opportunities and attend assiduously to this matter of integrity. There is nothing that so benefits a society as the competent, well-disposed public servant. And there is nothing that frustrates a nation's interests more thoroughly—or more insidiously—than public servants that have omitted to prepare themselves for the enormous demands that their offices will make.

We can now look at a few of the procedural and instrumental skills that the social service administrator might find valuable. Efforts in this direction begin by looking for attributes common to virtually all social service delivery systems, and, in the process, attempts will be made to lend immediacy and substance to some of the rather abstract arguments raised in this introductory chapter.

NOTES AND REFERENCES

[1] For an analysis of axiological predicates, see chapter four of John W. Sutherland, *Societal Systems: Methodology, Modelling and Management* (New York: Elsevier-North Holland, 1977).

[2] Indeed, Gunnar Myrdal has suggested that many social scientists are really more interested in changing the world than in understanding it. See his *Value in Social Theory* (London: Routledge & Kegan Paul, 1958).

[3] An explanation of these critically important, but seldom defined, referents is given by Gordon Rattray Taylor in *Rethink: A Paraprimitive Solution* (New York: E. P. Dutton, 1973).

[4] This is an essential message in the literate best seller, *The Crash of '79* by Paul Erdman.

[5] The distinction between the maintenance and developmental emphases, and the managerial prerequisites for the latter is discussed by Eric Trist in "Management and Organization Development in Public Enterprises and Government Agencies," prepared for the United Nations seminar on Use of Modern Management Techniques in Public Administration of Developing Countries. Available through the Management Science Center, Wharton School, University of Pennsylvania.

[6] For more on the distinction between semantics and syntax, see C. Wright Mills, "Grand Theory," *System, Change and Conflict* ed. Demereth and Peterson (New York: Free Press, 1967).

[7] For some idea of the implications of "benefits" see Lyden and Miller eds., *Planning Programming Budgeting: A Systems Approach to Management* (Chicago: Markham, 1972).

[8] The implications of accountability with respect to propriety (vis a vis performance) is given in chapter 3 of John W. Sutherland, *Administrative Decision Making: Extending the Bounds of Rationality* (New York: Van Nostrand Reinhold, 1977).

[9] Ibid.

2
DESIGNING THE DELIVERY SYSTEM

INTRODUCTION / Quentin Crisp, a wit of great repute, once suggested that: "The English think incompetence is the same thing as sincerity." To a great extent, Americans share this view. We tend to forgive errors of judgment when the errors are made by well-intentioned individuals; we tend to use the assertion that one did his best as a defense against virtually any type of mistake. All the arguments of the first chapter suggested that sincerity is not sufficient for proper management, and that integrity implies more than benevolent intentions. So we must now begin to look a bit more closely at the technological and intellectual aspects of integrity. The place to begin is at the point where the social service delivery system is designed.

The delivery system, after all, is the integral link between the resource system and the client or casualty populations. It, as a mechanism, is the residence for all social service management functions. The delivery system thus becomes the central focus of concern in this book: for the relative success of a social service program is usually determined at the point where the delivery system itself is designed.

So in this chapter, we shall look at the design process and take apart the prototypical delivery system by looking at its components from the broadest possible perspective. In subsequent chapters, each of the several components will be examined in much greater detail.

Discussion of delivery system design will draw on a health care delivery model which will serve as the case study throughout this volume; but the implications of the arguments raised in this chapter will, it is hoped, extend well beyond any single type of social service and well beyond any particular delivery mechanism.

A NOTE ON THE SYSTEM DESIGN PROCESS

Perhaps nowhere are the fields for consultancy so fertile as in the social service sector. Indeed, a significant proportion of the substance of social service delivery systems is owed to consulting firms, and not to the social service functionaries themselves. The majority of consulting complexes—and individual consultants—are both competent and reputable. As in any industry, however, there are the very few poor operators, and the social service manager had best be alert to some of the problems they can cause. Therefore, a deliberately dark and hypercritical portrait will be painted here of the type of consulting operation that can cause more difficulty than they resolve.[1] Again, such operators are encountered only rarely, but the havoc they can wreck is real enough. The problem for the social service manager about to let design contracts is this: these dysfunctional consulting complexes are often the smoothest, the best clothed and the most ingenious marketers. So the manager must expect that the symptoms of inadequacy—the warning signs—will be most subtle.

Most consulting operations begin with a technical entrepreneur whose ambitions cannot be contained by the government agency, the corporation or the university where he is employed. Success in various analytical or design projects may mark some individuals as distinctly more capable than others, and therefore give impetus to the formation of a proper consulting operation. In other cases, distinctly social or political factors may underlie the formation of a consulting complex. Particularly, the technical entrepreneur may use his government, corporate or university base to insinuate himself into various political circles, and attempt to trade on the equity of the position he holds rather than on the equity of his technical capabilities. The signature of the insinuative consultant is his attempt to personalize all professional relationships (a trick long practiced by producer-goods salesmen and capital-goods manufacturers). Inevitably, it seems, there comes a natural separation between the two functions of a consulting complex, marketing and contract performance. To most consulting executives, the focus of interest becomes the extroverted, dashing drama of the marketplace. It often happens that their energy is well invested at this level, for the granting of subsequent consulting contracts is not too often predicated on the quality of past performance . . . the government consulting market is, in a word, a place of most imperfect information. Thus, the extent to which there is a separation of marketing and technical functions is another signature of the consulting complex that may cause problems.

Those entrepreneurs who are particularly successful in exploiting their sociopolitical contacts eventually grow to impressive size and exude an aura of competence. Then they are in a position to exercise the most subtle and significant marketing leverage of all: to the extent that a government manager hires a *prestigious* firm to perform his technical operations, he is a priori immune from any repercussions should the job be done poorly or the design prove an embarrassment. This is the single most critical thing the consulting complex has going for it—the presumption that no procurement or contract officer could be blamed for hiring them should something go wrong. This selling strategy comes essentially from the tacit marketing strategy of large industrial producers. Let's, for example, say we are trying to sell computers and see how the strategy might work. Take a computer system manager who is asked by his organization's directors to develop a data processing system that yields the greatest output for the least cost, the effectively optimal system, as it were. Now, in all likelihood, an optimal system would involve the interconnection of different components from many different manufacturers, for no single vendor has the best components for all functions. This computer system manager is now in a very vulnerable position, for what if the collection of different manufacturers' components turns out to have problems? In defense of his decision he may suggest that every system installation has problems and that all he did was follow instructions: he researched the market and put together what on paper appeared to be the optimal system. But he is in a very precarious position, for none of these smaller individual manufacturers will be there to share the blame, and the manager's superiors will in all likelihood not share his technical sophistication. The safer and easier strategy for the manager, then, is this: buy an entire system from a single large vendor. If anything goes wrong, the large vendor takes the blame, for no senior official could possibly fault the system manager's judgment for having hired an undisputed leader in the field. Moreover, the manager who buys a ready-made system from a single producer is saved the trouble of having to do any original research.

This is perhaps a somewhat cruel assessment of producer-goods marketing, yet the lesson is clear. The manager who chooses to deal with a prestigious contractor (a) is de facto relieved from responsibility for any errors; (b) is relieved from the necessity to do any original technical assessments; (c) borrows some of the prestige and moment of the large entity which comes to court him; and (d) benefits from the apparent esteem and gratitude (often expressed in the most tangible way) by the heavyweight marketing men of the corporate giant. But, and this is virtually always the case, the organization whose interests the system manager is there to protect pays a premium for the manager's expedience. This premium is tacit; it doesn't appear as a tangible opportunity loss on the organization's year-end financial statements, and therefore generally never becomes a real factor in the contract decision process. But it is there, for the contractor will pass on the costs of doing the manager's job and also pass on costs of any system suboptimalities in forms of operational inefficiencies.

There is a significant problem with this process, one which is seldom made explicit. When the public administrator moves to hire a large prestigious consulting entity to perform his system design functions, he may think he's getting a whole slew of Ph.D.s and elegant scholars to do the work. But, in fact, the highly credentialed representatives of the organization may be just the front men; what he often gets, instead, is an ill-articulated collection of transient M.B.A.s or low-level engineers to do the actual work with the consulting organization itself being little more than a glossy marketing cadre. In short, many consulting complexes are little more than "body shops," acting as brokers for unemployed technicians. Even when the consulting complex is a substantial, stable one with a permanent staff, there can be problems. Established consultants often sell precedent; nothing makes a consulting operation as happy as being able to sell a contract to one party which has already been performed for another party. When reviewing reports by the large consulting operations, they tend to look alike, even when they deal with different functions. This is because the consulting firm has an inventory of *boiler plate* (information or data, etc., left over from some previous job) that with a little imagination can be applied to the new job and thereby earn another fee.

Now there is another critical problem with the institutionalization of government contract research. Fundamentally, the welfare of any firm in the business of consulting depends on being able to maintain the affection of the contracting and procurement officers themselves. The unwritten rule, then, is to be sure that nothing in the final project report can cause existing functionaries any embarrassment, or cause irritation among individuals who might someday be in the position of throwing some business in the firm's direction. It should be no surprise, then, that projects that were originally intended to reform an agency or reorganize a function so often turn out to be completely lacking in content or substance. Reformational or reorganizational projects, while lucrative, are also perilous; reforms always offend somebody, and, given the structure of competition, most consulting operations cannot afford to offend anybody.

In terms of actual technical process, the consulting firm is always under pressure to cut corners. Boiler plate, low-paid operatives, minimization of original research, etc., are powerful tendencies in the consulting business. Even nonprofit entities (and some university-based consulting complexes) are prone to minimize the substance that goes into a project relative to the return (the contract price). The excess funds may then be translated into new professorial or research lines or into the conspicuous amenities or infrastructure that most think tanks seem to favor. When we combine the venality of some consulting operations with the sad lack of mature talent among the hirelings who do most of the actual work, we get the inevitable result: very expensive, time-consuming, highly touted projects which produce the most pedestrian, sophomoric and "safe" prescriptions.

What then are the alternatives for the public administrators who need system design or analysis work done? There are basically four:

1. Have the work done by a smaller, newer consulting operation that is still trying to make its reputation, and make sure the contract specifies which individuals are actually to do the work over what intervals.

2. Make the R.F.P. (Request For Proposal) process really work for you. As things now stand, the large consulting operations often ingratiate themselves with the bureaucrat by offering to write the R.F.P. for the agency; often, the resultant R.F.P.

emerges in a form that suggests that the company who wrote it is the *only* company competent to actually perform the work. Therefore, the R.F.P. process becomes a stunning sham rather than the source of efficiencies it was originally intended to be, and the competitive bidding process is subverted at its initiation.

3. Hire individual consultants who have some other source of income. Particularly attractive, in this respect, would be academics who are able to take a leave from a university and who are protected by tenure or other sources of job security. The chances are that these individuals will be less afraid to report criticisms than the consultants whose entire income depends on their being able to avoid conflict or recrimination. They are individually accountable for their work and carry no overhead costs.

4. Hire in-house people capable of performing system analysis and design operations in their own right. More and more of these people are being trained in system and business administration programs and are available to the public sector. In the social service sector, at least, system analysis and design activities must be the rule rather than the exception, that is, they are processes that should be conducted with consistency and constancy. Therefore, in a properly managed agency the system people are unlikely to run out of useful work. Because of their intimate association with the functions of the agency, they are likely to be more involved with the substance of the agency than would the itinerant operatives the consulting operation might put to work.

Given some consideration as to who might best do system analysis and design work, here are a few suggestions about the way in which the work might be performed. Within the context of the social service sector, a "system" may be taken to read *delivery system:* the mechanism which husbands the resources associated with some particular social program or some set of programs. The system design and analysis process, in this context, thus means explicitly the design of delivery systems. And with respect to this process, a rather considerable experience in system design exercises (not all of them successful, by the way) has taught me the following:

1. Never begin a system design project by collecting data or doing field research. This is the favorite ploy of the professional consulting complex, as it puts lots of relatively mediocre people to work doing lots of mechanical processing for which high fees can be charged. As a rule, the data collected at the initiation of a design process is useless, or at least some significant portion of it. The reason is this: in the social service sector, particularly, there exist very few successful precedents from which adequately promising prescriptions may be drawn. Therefore, there really is no point in paying people to find out what other people are doing. Nor is it "scientific" to try to derive hypotheses from data, not unless the subject being studied is a relatively simple one which is amenable to laboratory manipulation (which is simply not the case with social service issues). The guiding question the system designer must always ask himself about data is this: *Compared to what?* Doing field research to find out what others are doing (unless there is indisputable evidence that they are performing successfully) always leaves this question unanswered, for there is no normative reference against which to compare the information generated about current processes.

2. It is suggested that the system design process in the social service sector begin with a *normative analysis* and only subsequently engage in field research. Here's the rationale. The vast majority of delivery systems operating in this and other developed nations are ad hoc enterprises, that is, they just evolved; they were not really designed. Moreover, the technology of system design and social service management is a fluid subject, constantly changing. The ad hoc systems generally employ only the most limited technology, much of it quite immature. Therefore, there is really not too much positive information to be gained by reviewing existing processes. Moreover, when the design process begins with field research, there is the almost inevitable bias of precedent which sneaks in: the existing forms rather than the optimal forms become the often explicit (but sometimes tacit) referents of the designers, and the cause of true innovation and reform is frustrated. Therefore, the first phase of the design process should be to develop a normative model, setting out the best of all possible worlds as nearly as the analysts can

deduce it.* That is, the initial referent for the delivery system design exercise is a product of thinking, not a product of data collection.

3. When system designers are forced to operate in the normative mode, they are required to be original and innovative. Therefore, it is often important to conduct the normative analysis using people who are more distinguished by their intellectual creativity than by their experience with a particular function or social service context. That is, the normative model is an analytical, almost academic exercise. Many students of system science are now being trained to perform deductive analysis exercises,† so it is not impossible to find people who can do this job. The advantages, again, are these: (a) there is no opportunity (or at least reduced probability) of having the designers simply reinvent the wheel by taking existing and suboptimal systems and reconstituting them; (b) the *idealized* criteria for the delivery system become explicit at the very start of the design process, and thus become the major foci for subsequent analysis (as will be shown shortly); at the very first stages of the design process, the optimal criteria become explicit, whereas they may never appear at all in traditional field-research-oriented design processes; (c) where the delivery system is to be unprecedented—serve a function that does not now exist—then it may be possible to avoid the expensive field-research and data-collection activities altogether, thus accelerating the time to implementation.‡

4. Even where the mission for the system design process is to implement a service already existing elsewhere—or to reform an existing process—the normative model has the following advantage: it tells where to direct subsequent field research and data collection activities. That is, the normative model now expresses the desires or optimal criteria of the system architects or contractors; the properties of existing systems would then represent the base reference.

*Using a priori, intuitive, theoretical, etc., referents as the basis for the deductions.
†It should be noted that most of the functionaries employed by the large consulting operations do not have training in this highly specific technique, but tend rather to be drawn from relatively traditional engineering or business administration curricula.
‡Field research will, of course, be conducted during the implementation phase as a necessary process for the development of data bases and the attuning of the system to local conditions, etc.

Therefore, we now have the answer to the question asked earlier: Compared to what? Compare the existing system to the normative system, and the qualitative (and sometimes quantitative) differences are the focus of subsequent design activities. In effect, we are now prepared to deal with this question: How do we go about transforming the existing system into the normative system? The ancillary question then is always this: How feasible is the normative referent given economic, social, political or technical constraints? How close to the ideal can we really expect to come? In almost every case, the resources that are saved by not starting to willy-nilly collect data will mean that there are more funds available for the implementation function and a higher probability of getting closer to effective optimality than would the case be were the normative model not made the first order of business.

5. Given the normative referent and the existing system, the task then is to involve the existing agency personnel in the concretization and detailing of the system design. Questions about the feasibility of the normative criteria may in part be answered by the agency's operating personnel, and their particular specializations (and a priori informational sets) may be drawn on to flesh out the normative model. At any rate, it is important *not* to involve the existing agency functionaries in the normative design process, for there is always the danger that the subsequent system design will merely reflect the limitations of the existing personnel. This is always a problem for the professional consulting complex, for it cannot afford to alienate the agency operatives or appear to bypass them. Therefore, even at the expense of optimality or objective system effectiveness and efficiency, consultants may often let the existing staff dictate the limitations of the system design so that they are not threatened. The result, again, is a "reformed" system that has many of the defects of the original system. Of course, the problem of how to keep staff limitations from restricting the normative design is obviated when the project is aimed at developing a brand new delivery system. Even then, however, agencies often hire a delivery system manager *before* a design is delivered. He either dictates portions of the design to reflect his own predilections, and thereby dilutes the normative compo-

nent; or, as is often the case, the director will simply not have capabilities congruent with the implications of the design, causing a built-in weakness to be dealt with at the implementation stage. Therefore, it is most strongly recommended that an agency contract for the system design to be completed *before* it seeks a director and personnel to actually implement the system. In this way the normative criteria of the system design determines the attributes of the personnel to staff the system, and not the other way round.

These general assertions can perhaps be lent some substance for the reader by the presentation of an example of what a normative system design looks like and its implications for actually administering the design process itself.

THE NORMATIVE DESIGN CRITERIA

As was suggested in the preface, we are going to be working our way through an actual system, a model of a health care delivery system that has certain definite criteria it is to meet. These criteria, established for the most part prior to the initiation of the delivery system design exercise, really represent the normative components mentioned above. The term *objectives* will be reserved for those operational system goals that have a quantitative component. So, when speaking of normative criteria, we are really talking about setting out the fundamental qualitative properties the delivery system is to reflect.

For the most part, given the currently immature state-of-the-art of delivery system design technology, the normative criteria will look pretty much alike for all delivery system contexts. Therefore, those employed for the health care delivery system (the HCDS) will probably be of interest to those concerned with nonmedical social service programs as well. At any rate, the next few pages will report on the nature of normative criteria, and the criteria defined will have quite broad an application.

Effectiveness, Efficiency and Dignity

Initially, it is necessary to specify what is called the *enveloping criterion* for the delivery system. This is a criterion which serves to distinguish social service systems from other types of systems.

The point, then, is that virtually all social service systems will have the same basic strategic criterion. That which seems most appropriate, given the discussion of integrity in the first chapter, is the following: *Any social service system should be designed to operate in the neighborhood which defines a favorable balance between internal efficiency, contextual effectiveness and client dignity.* This is simply a restatement of a strategic dictate introduced earlier: that social service managers seek always the best balance between compassion and competence. In broader terms, it implies that the social service program always operates in a crosscurrent between economic and social, and local and universal constraints.

However, were we talking about the development of an organization to pursue some ambition in the commercial or industrial sector, a more appropriate criterion (at least from the parochial economic standpoint) would be to maximize profits, or maximize rate of return on invested capital, etc. Therefore, from an operational perspective, the sufficient dictate for the commercial or industrial manager is efficiency. Efficiency enables the competitive enterprise to produce a given product or service at the price which (approximately) maximizes demand for its output, given some elasticity factor. Efficiency criteria also implies that the firm will produce at a level which maximizes the excess of outputs over inputs, which means that it will generate a level of aggregate output which is neither too large nor too small and that the firm seek, and not exceed, the *optimal scale of plant.* In computational exercises, this is the point where marginal revenue (incremental, pro rata profit contributions) exactly equals marginal cost (incremental costs associated with the production of the n'th item of output). For the most part such calculations are impossible to make for most commercial firms, so managers will content themselves with a "neighborhood" or approximately optimal position. Where economic and technical skills are scarce among a firm's management personnel, a cruder and reflective criterion will be used, perhaps maximal growth or largest share of market, etc. These *reflective* criteria merely establish a ranking for a firm with respect to its competitors, and do not of themselves tell anything about the economic optimality of the enterprise. Rather, they simply suggest the ordinal or relative efficiency of one firm compared to other firms in essentially the same business, e.g., members of an oligopolistic or purely competitive industry with little product differentiation.

It goes without saying that reflective performance criteria are often denied organizations operating in the social service sector. It is very unusual to find two such organizations that have exactly the same missions or are resident in exactly the same environmental contexts (though inadvertent redundancy is not uncommon). Therefore, it does not usually make much sense to talk about one hospital being reflectively more efficient than another, or about one school being a better social investment than another.

Yet, for purposes of the rational allocation of public resources, such comparisons must be made. Strict economic criteria must be qualified by the use of social or contextual equivocations. For example, one might suggest that the basic business of most welfare programs is simply to deliver a check from the public treasury to a member of the casualty system, the welfare client. Now, were we to employ a strict efficiency (economic) criterion, we would ask that each welfare program act to *minimize transaction costs.* In effect, this would mean that each program would seek to actually transfer the funds with the lowest associated administrative overhead. How might this be done? For one thing, minimization of the staff associated with the welfare office itself by having a computer do as much processing as possible, and discouragement of contact between agency personnel and welfare clients, for face-to-face contact (the function performed by case workers, per se) are costly adventures. To minimize the costs of auditing for fraudulent payments or unqualified participants, have the computer be in contact with other computers, possibly those of the IRS, the unemployment insurance office, etc. In short, automate, as completely as possible, all phases of the welfare program. If reflective criteria were employed to compare the efficiency of several welfare programs (say in different neighborhoods or operating among different client populations), we would simply look at the following ratio: T/A—the ratio of total funds transferred relative to administrative overhead. On this basis, the agency which was able to get rid of the most money at the least cost would have proven itself the most efficient organization.

What in practice defeats such measurements is this: efficiency and effectiveness in the social service sector may often be competitive ends. Particularly, if one defines the ambitions of a welfare program to be something more than a transfer function (e.g., getting money from one segment of the population to another), then the strict economic criterion does not *exhaust* the relative performance criteria for the welfare program. As it happens, most welfare administrators—and certainly the line personnel of those agencies—do have ambitions that go beyond the mere maintenance of their clients. Now it may be that for various reasons the developmental impact of existing welfare programs is very low (and often insignificant). Still, most case workers and welfare professionals would agree that part of their mission is to provide their clients with more than money: to provide them with guidance, comfort, direction, sympathy, purpose and, in some cases, a motivation to better their lot. Therefore, given this range of mission orientations, it would not be appropriate to attempt to measure the performance of a welfare program merely on the basis of its transaction costs.

In more specific terms, the conflict between efficiency and effectiveness comes always to haunt the social service sector. Let's take a simple example. Suppose that an agency is dealing with a casualty subpopulation in some urban area, and that the legislated mission is to get resources (shelter, food, clothing, medical care, etc.) to these clients to provide for a subsistence that is adequate to needs. But suppose the agency was restricted just to the transfer function. This would mean that the task begins and ends with seeing that the client gets a check every month which, on the basis of some strategic and generalized calculation, appears to provide adequate subsistence. In short, some bureaucrat somewhere has determined that a family of four persons needs $X per month to provide adequately for itself (with adequacy also being determined centrally or generically for some region). But what is the central assumption behind the determination that $X is a sufficient level of support? The assumption is that this family of four social casualties will be able to reproduce the decision process that the central decision maker went through. That is, the central decision maker set a level of support on the basis of his determination that of the $X so much would go for rent, so much for food, so much for clothing and medical care, etc., given a structure of prevailing market rates. In all probability, occupying as he does a relatively high position in some bureaucracy, this decision maker may be thought to be an intelligent consumer. That is, he has the judgment and experience to make proper, well-schooled consumer decisions.

In many cases, however, the social casualty will not be a well-schooled decision maker in so far as his consumption or expenditure problems are concerned. For example, he may be very gullible about paying rents, in short, a ready-made victim for the slum lords. Moreover, the social casualty may be very susceptible to medical quacks or to the prescription mongerers who make a habit of exploiting welfare or Medicaid-Medicare clients. And as suggested earlier, the social casualty will, in all probability, shop for his food and clothing in that inflated ghetto economy where prices for basic commodities may actually exceed those of a prestigious neighborhood. Therefore, when considering the consumer innocence of most social casualties, there is some likelihood that any centralized determination as to dollar requirements will be insufficient in practice. In fact, we could expect that $X will not provide the basic level of support the bureaucrat (or the enabling legislation) intended. Therefore, the mere transfer of resources does not even meet the legislated economic ambition of the welfare program: to deliver to client populations a monthly donation sufficient for some minimal level of sustenance.

To the extent that this ambition is not met, the welfare program would be *ineffective*. But on the surface, the welfare program could still be *efficient*. That is, it could be very clever and expedient in delivering the monthly check to its clients. As is obvious, though, the most important performance measure for any organization is to be effective first, and only then to be efficient, for efficiency must be measured as the cost associated with obtaining a certain level of effectiveness. In the above example, the tacit criterion for effectiveness was this: to assure minimal adequate sustenance for social casualties. The operative strategy, taken from the strict mechanical or economic perspective, was to seek this effect through the simple transfer of resources, with the subordinate criterion of efficiency being to minimize the cost of transferring these resources. But now there is the hitch: the desired effect of the program cannot be achieved because of the innocence of the social casualty as a consumer. Therefore, the program is *ineffective*, and there can be no applause whatsoever for its imputed efficiency. In short, in an effort to maximize efficiency, the agency became ineffective.

What, then, would the welfare professional have to say about this? He would probably suggest this: in order for any welfare program to be effective, it would have to make available the necessary contacts that would attempt to protect the social casualty against the adverse affects of his lack of acumen as a consumer. That is, case workers would have to be available—in personal interchange with the social casualty—to provide him with decision guidance in the selection of a place to live, in the preparation of a food and clothing budget, in the selection of medical providers, in the matter of going into consumer debt for furniture, etc. Moreover, these face-to-face contacts would have to be used to make the social casualty a well-informed welfare consumer, alerting him to all the various social services for which he might qualify. Now, these expensive, time-consuming contacts add considerably to the administrative overhead of the agency, making the T/A ratio much lower than would be the case with the agency which just sent a check but offered no subsequent guidance or consultation. But the central and telling argument is this: these contacts are not luxuries but are *necessities* if the welfare program is to be effective . . . if, in fact, it is to meet its basic mission.

Let's broaden the context a bit. In the late 1960s, an agency called the OMBE (Office of Minority Business Enterprise) was formed as an adjunct to the Department of Commerce. The ambition was a noble one: to stimulate the development of businesses owned by and serving minority populations. Resources were made available to the agency for transfer to potential minority businessmen. Now here is a situation similar to that facing the prototypical welfare agency. The administrative overhead of the agency may be kept low to the extent that the agency sees itself simply as a conduit for funds. In such a case, the way to minimize transaction costs is as follows: (a) keep agency personnel to a minimum; (b) keep program promotion to a minimum; (c) give a few large loans rather than a lot of little loans; (d) minimize agency contact with recipients of the funds. A set of operational policies like this might, at first blush, give evidence to the agency head's appreciation of efficiency and to his vision of himself as a hard-headed businessman in his own right. Look again at the criteria of effectiveness (the mission) for the OMBE: to stimulate minority business enterprise. Here we find the basic contradiction between the conditions of effectiveness and efficiency. Particularly, efficiency dictated a minimization of contact with clients and with ancillary expenditures, but the very nature of the minority community would suggest that there were not going to be large numbers of qualified businessmen there to receive the loans or make proper use of them if obtained. Therefore, the criteria of effectiveness would imply additional aspects to the program, each of which would increase the agency's transaction costs: (a) There would have to be an effort made to stimulate a class of minority entrepreneurs; that is, a considerable and sophisticated promotional program would have to be launched to develop an entrepreneurial motivation among historically deprived populations; (b) Next, for those entrepreneurs who did express a demand for agency support, there would have to be a program to teach them how to use funds or to make sure that continuing consultative support was available to them. In actual practice, neither of these ancillary programs was adequately supported, and the net impact of the OMBE was most undramatic. It tended to result in the clustering of power among the few minority entrepreneurs which existed prior to the program's inauguration and did not do much to expand the population of minority businessmen. Moreover, those novice entrepreneurs who did receive funds had a staggeringly high rate of bankruptcy. The reason was simple. Banks and other lending institutions were given funds to relay to minority businessmen. These funds carried, so far as the banks were concerned, a zero-opportunity cost; that is, they were either delivered to minority businessmen or foregone altogether. Therefore, the banks were not motivated to be very selective in the investments that were made (as they were virtually insured against possible losses from default or bankruptcy, etc.). So, once they made a loan, they were not tempted to follow it up by providing business advice or professional consultation to the new businessman. Were they to make available business advice, this would have resulted in a positive opportunity cost for the banks;* that is, their personnel could have been used elsewhere in a more favorable or more potentially profitable situation (providing advice to larger customers or to those who have uninsured loans, such that the threat of default is to be avoided as completely as possible). In summary, then, much of the failure of the OMBE program can be attributed to the fact that the agency management considered themselves performing a simple transfer function and were naive about the necessity for first of all developing a minority entrepreneurial class and then educating it—and guiding it—in the wise use of the resources made available. In a sense, then, any efficiency which the OMBE exhibited as a conduit was more than

*This criticism is generally not true of those OMBE outlets associated with university-based business schools or other nonprofit sponsors.

offset by its ineffectiveness as a real developmental program.

Another important example comes to mind (one which will have more said about it in later sections). This is the matter of a national health care program. A look at the vast majority of schemes that have been proposed shows virtually all of them having one property in common: they tend to rely entirely on some type of insurance or prepaid modality. That is, everybody to be covered is to be given some sort of Blue Cross card or its equivalent. Arguments among the advocates of the different schemes are thus not on the matter of a central strategy, but rather on distinctly tactical issues: Who is to be covered? Is coverage to be voluntary? What medical conditions are to be supported? But in light of what was just said, the prepaid or insurance modality may itself be in doubt, and therefore a national health care program based on this strategy may be foredoomed to ineffectiveness. Again, we have the critical qualification: some of the medically indigent in this country will probably not be equipped to make rational decisions about matters of health care. They will, again, be susceptible to charlatans or quacks; they will perhaps neglect preventative care and perhaps be even casual about the systematic treatment of chronic conditions; they may be ignorant of the difference between various types of practitioners, and be generally unaware of the nature of various medical specialties, etc. Now, admittedly, the prepaid (insurance) modality does appear to have advantages from the perspective of efficiency; that is, it is proposed as being the way to minimize transaction costs. Unless the nature of the client is taken into consideration and his need for positive guidance in decision making, then efficiency might be obtained only at the expense of real effectiveness, and another social service failure might emerge. In short, advice and direction must accompany the funds in ways which will be made clear later.

It is thus critically important to realize that in the social service sector effectiveness and efficiency are not the same thing, and this is one of the major operational differences between social service and commercial or industrial enterprises. In the commercial-industrial domain, the effectiveness of a firm is usually looked at with respect to the dollar return it provides; between two firms in a competitive relationship, *relative returns are directly related to relative efficiency.* But in the social service sector, this simple relationship does not hold true.

Therefore, let's look again at the generic (enveloping) performance dictate for the social service program: *it should seek the most favorable tradeoff between economic efficiency, strategic effectiveness and client dignity*. The first of these criteria simply reflects the responsibility of the social service sector to those who provide the resources with which it operates. The effectiveness criteria simply warns the social service administrator against being overly concerned with the economic criteria, and alerts him to the fact that the intellectual demands he will face may often be more complicated than those faced by industrial or commercial managers. If he becomes too enamored with the efficiency of his enterprise—thinking of himself as too much of a businessman—he may guarantee the failure of his program. By the same token, if he is unconcerned with any efficiency or economic constraints, then he may purchase a certain level of effectiveness at a much higher than optimal price, and therefore be condemnable as a fellow who contributed to the waste or erosion of the real asset base of the larger society.

Finally, there is the matter of client dignity, the element of the compassionate aspect of operations in the social service sector. This criterion simply asks that the social service manager consider the mechanics of his system from the client's standpoint. In a sense, this means defining what the term client really means. It means, first of all, that the social service functionary has a professional relationship with his client; that is, he is interested in his client as a person and is duty-bound to consider at all points that client's complex welfare. Therefore, in actual practice the social service functionary has to walk the always fine line between being the client's parent and his mentor. The parent is expectedly sympathetic and protective; yet in the role of mentor, it is the social service functionary's responsibility to make the client responsible—to the extent possible—for his own welfare. That means that the client-oriented objective of every social service functionary should be to put himself out of business by developing the client to the point where the interjections of the social service worker becomes gratuitous. In operational terms, the criterion of dignity then demands that requirements are not imposed on the client which are arbitrary or inconsiderate (e.g., making him come in for a personal visit every time he has a procedural question or making him constantly requalify for benefits even when he is unlikely ever to be disqualified), but also that there is not a failure to make demands on the client which, if met, will be to his benefit. Later discussion of the HCDS case study will have more to say about dignity and the client-professional relationship.

The Technical Management Criteria

It is useful here to say a few words about the way in which one might realistically expect to locate a most favorable tradeoff between efficiency and effectiveness. For the moment, leave aside the third criterion, client dignity; this, as will be shown later, is determined solely by context and the axiological predicates held by the social service staff itself (though sometimes it can be affected by a militant client population). At any rate, this matter of the effectiveness–efficiency tradeoff takes us squarely into what might be called the realm of technical argument. It is here that the intellectual demands on the social service manager become most compelling.

These next few pages may be of little interest to the casual reader, and even to some of those who have a definite interest in the arguments, there may be some miscomprehension. But the purpose here is to explain what *should* be within the capabilities of the social service manager. For those who wish to act upon this advice, a reading of a standard microeconomics text and a review of some of the fundamental literature of operations research (or management science) is suggested. Or, as an alternative, the reader might want to enlist the aid of a friend or associate who has some knowledge of these technical fields. The interest here is not in showing the reader how to formulate an optimal tradeoff; rather, it is in letting him know that it can and should be done, and that in some cases it is a good investment to hire an expert in such things if one himself does not wish to acquire the facility.[2]

Let's begin, initially, by suggesting that there is a *ceiling level of effectiveness* for any social service program and that this ceiling level of effectiveness is determined by a simple ratio: X/Y. In this ratio, X indicates the resources (e.g., funds) the program has available to it, and Y indicates the size of the client population over which those resources must be distributed. The proper name for this ratio is *the per capita resource level,* and the strategic implication is this: in the usual social service context, the higher this ratio, the greater the level of service one can offer to the average client. The corollary is this: in the social service sector, effectiveness of a program is generally positively (and, within limits) directly correlated with the level of service we are able to provide. That is, an educational program, all other things being equal, has a higher *potential* effectiveness the greater the resources available per pupil; assuming that internal allocations are made rationally, a higher per capita resource level can provide a lower student–teacher ratio, better and more materials per pupil, improved lighting, better designed classroom facilities, better recreational facilities, meals with greater nutritional value, etc. Or to take another example, a medical clinic's effectiveness is related to its per capita resource level in the following way: a higher per capita resource level implies a better ratio of doctors to patients, more ability to provide preventative rather than just acute (or crisis) care, better and more modern diagnostic equipment, etc.

Therefore, if there is a central strategy for the senior executives of any social service program, it is this: that they act to obtain a steady (secular) increase in the X/Y ratio for their organization, for it is only in this way that they can act to increase the *potential* effectiveness of the program. Therefore, the effective social service manager—looked at from a strictly local standpoint—is one who assures that $(X/Y)_t > (X/Y)_{t-1}$ . . . that is, that during any interval, he has been able to achieve a situation where the current *(t)* per capita resource level is higher than that which was obtained at the previous *(t−1)* period.

When personally counseling executives of a social service agency, they usually agree that this objective is a valid one. But they then ask the crucial question: How does one assure the secular increase in the per capita resource level? There are, of course, several ways. First, he may seek to broaden his funding base—that he try to *maximize unallocated resources* available to his organization.* He does this by seeking a noncategorical, nonformula-based source of funding. This is important because the usual method of financing social service operations is to tie them to a categorical or formula-based budget. This means that funding is assigned on the basis of the number of clients the organization must serve. The nature of this funding means that resources increase or decrease only proportionally with the increase or

*For a more complete analysis of this dictate, see chapter 3 of John W. Sutherland, *Administrative Decision Making: Extending the Bounds of Rationality* (New York: Van Nostrand Reinhold, 1977). For agencies or programs funded on a categorical basis, this often means that the program manager must seek to increase the proportion of overhead associated with his funding flows. This overhead may then be translated into directly productive funding at the local level, even though it was intended by the central funding agency to support accounting, reporting and other nonproductive activities.

decrease in client population, that is, the agency gets so much per client and no more. This, then, sets a definite and intractable limit on the per capita resource base, and in fact then sets the potential level of effectiveness at the minimally legislated level. Therefore, the first task of the service manager is to be constantly alert to opportunities for attracting noncategorical funds, for it is these which may be used at his discretion to amplify the level of service the agency can provide any client population. In practice, this is equivalent to demanding that the social service manager think of himself as more than a bureaucrat; in fact, it means that he must cultivate the alertness, energy and imagination expected of an entrepreneur. As is to be expected, this alternative is not very attractive to most bureaucrats, as it introduces an element of anxiety and assiduity that most public servants are perhaps predisposed to avoid.

The second strategy for increasing the per capita resource base for any social service program is equally demanding. This consists in mounting a sophisticated lobbying effort to have the formula budget or categorical funding base revised upward. This means that the bureaucrat must now become a politician of the most constant application. Again, this is perhaps a role which is fundamentally unappealing to the usual social service function-ary.

The third strategy is even more complicated. It involves a two-pronged attack. First, it is necessary to have the formula or categorical budgeting mechanism translated into a "block" mechanism, one where the agency is given a block of funds to work with over a period, irrespective of the actual client popula-tion. Now, to turn this to the agency and the client's advantage, it is necessary to then make every attempt to reduce the client population relative to this block of resources. This strategy, of course, would be most appropriate to a developmental program which is anxious to minimize rather than maximize the client population. It is also appropriate in those instances where a *triage* strategy is dictated, where the resources must be distributed asymmetrically across a population if the program is to have any real effectiveness (as the formula or categorical budget level is so low as to be able to help no one if the resources were distributed equally to all clients).

When considering that the formula budgeting or categorical budget levels associated with most social service programs are indeed set at a minimal level, it becomes imperative that the social service manager seek to revise his funding base through one or another of the mechanisms just indicated. Therefore, the role of social service administrator is not for the lazy, the faint of heart or the unimaginative. To be sure that one's program is indeed *effective* requires not only considerable substantive skill and knowledge (that is, a realization of the nature of the client population and a strong technical background in the kind of service to be delivered), but also considerable political or entrepreneurial capability.

Now there is also that matter of managerial integrity outlined earlier; it requires the social service administrator to know when enough is enough . . . when his per capita resource base should not be increased any further. What this means is that the social service manager should seek to increase the resource base of his program only to a certain point (and in some cases, as will be shown shortly, he should refuse to inaugurate a program altogether). This makes sense when looking not at an individual social service program, but at the social service sector as a whole. If there were a central authority—a grand manager, as it were —responsible for the integrity of the aggregate social service investments, his responsibility would be to allocate resources in such a way that each investment of the public funds carried an effectively zero-opportunity cost. Again, a zero-opportunity cost associated with some allocation of resources means that there is no other investment which might have been made which would have resulted in a greater return (or a higher level of marginal benefits or incremental effect, etc.). Because we do not really have this grand manager, the effective authority over level of funding falls on the individual program manager (or in some cases, the executive in charge of some collection of service organizations). Now it may be that most managers seem always to be complaining that their resource level is too low relative to the tasks that must be performed, and for the most part they might be right about this. But it is possible to take a very broad look at the matter of effectiveness and efficiency and consider at least four different situations that might present themselves to the system manager. In this respect, consider Figure 2.1.

Functions of this type are a frequent fixture in microeconomic analyses (which may generally be said to supply the theoretical rationale for most management science instruments). Basically, the curve suggests that a given increment of input (resources) will yield variable increments of output, depending

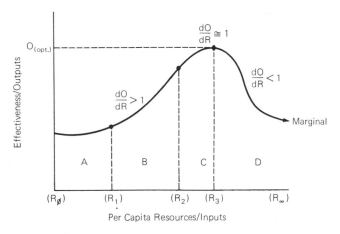

FIGURE 2.1 / The generic production function

on when the input increment is applied. The "when" here refers to the cummulative level of inputs, measured along the X axis. The morphology or shape of the production function is not fixed for any given enterprise, but may be changed by technical (processual) or managerial innovations. In general, however, for any given management and production strategy, such curves can yield what is called the optimal scale of plant, the size of enterprise the organization should not exceed if it is to remain economically optimal. It is thus as applicable—in concept—to social service systems as to commercial manufacturers.

In Figure 2.1 it is the interpretation of each of the four lettered areas that is of concern:

Area A: Even though we are increasing the level of inputs (from $R_\emptyset$ to R_1), we are not getting any real effect for those dollars, that is, the level of output or effect remains virtually constant throughout that range. The reason is that the level of inputs is simply not sufficient to produce any positive impact. Simply, a *critical mass* has not yet been reached. For our purposes, the implication is that the prevailing per capita resource level (X/Y for the range $R_\emptyset - R_1$) is so low that no one benefits. In this case, given the assistance available to him, the client is really no better off than he was when no resources were available. In such a case, the expended resources are entirely wasted, and the responsibility would devolve on the system manager to refuse to inaugurate the program at all.

Area B: In this range, beginning with resource (or input) level R_1 and extending through R_2, there is a much more favorable situation. For here, each time the incremental level of resources is increased, there is a *more than proportional increase* in effectiveness (or output). In technical terms, this means that the relationship between inputs and outputs is positive and accelerating (as indicated by the fact that the first derivative associated with the B area is greater than unity). Now, economic theory will say that even though there is that acceleration of effect in this range, it still represents a suboptimal region, for until reaching point R_2 (associated with the optimal output level indicated on the vertical axis), any increase in investment will prove to realize an even greater incremental benefit or output margin. Therefore, the program manager should continue to demand inputs up to the level R_2.

Area C: This is the *neighborhood of the optimum,* and sets the effective limit on the level of resources which should be allocated to the program. In this range, inputs and outputs tend to be proportional, such that we get out (in terms of effect) just about what was put in. At this point a limit should be put on the resources willing to be devoted to the program.

Area D: Beyond resource or input level R_3, is the domain of decreasing returns to scale, that is, getting back, in terms of effect, *proportionally less* than was put in. Therefore, further investment in this program becomes uneconomical and a misallocation of scarce social service resources.

Because this theoretical construct is so critical to a comprehension of the technological aspects of social service management, let's spend a few words on its practical implications. Just by way of example, suppose that an educational agency has responsibility for some training function. Their first concern, as suggested earlier, is with the relationship between level of service and system or program effectiveness. Within this context, the relationship is fairly simple to suggest: the more educational resources which can be devoted to each pupil, up to a point, the greater the level of educational effectiveness we may presume to obtain. With this relationship in mind, some practical import can be given to the four regions. In the range defined by area A, the implication is that virtually no effect for the resources committed can be expected. For example, suppose that there is a difficult subject to teach, one commanding much personal attention from the instructor, but resources that permit only a student-faculty ratio of 400 to 1. In such a case, the educational effort will be so hampered as to make it gratuitous. Only the most naive manager would agree to mount such a program. Now, this simple example clearly shows the concatenated implications of raising the resource or input levels. As more resources are added, the student-teacher ratio gradually declines to the point where it reaches an optimum somewhere in area C. Now, if further additions to the teaching staff continue, we may eventually reach the point of diminishing returns, where the addition of faculty—for a given student body—can actually produce negative results. (Suppose there are two or more teachers each trying to command the attention and interest of each student simultaneously; the result would be virtual chaos and an actual negation of educational effectiveness.) Or, beyond some point (R_3), addition of faculty will produce a greatly dampened effect . . . the system is approaching *saturation*.

The point, then, is this: it is incumbent upon the social service manager to determine, through analysis and experimentation, the optimum neighborhood for per capita resource levels. He should be bold and determined about increasing resources to that level, for the basic effectiveness of his program—and his responsibility to his clients—is served thereby. He should also be alert and responsible enough to put a ceiling on the resource base when he sees he is moving beyond the optimal neighborhood into the domain of diminishing returns. Surely, the reader can see that this function might better be performed by some centralized, higher-level decision authority. To be able to watch over several or many programs—to chart their progress along the generic production function—would demand an individual of considerable technical skill. To the extent that such individuals are going to be hard to find, it is incumbent upon those technicians who undertake to develop delivery systems to see that the system itself inheres the capability to assist the nontechnical manager in determining the optimal tradeoff between effectiveness and efficiency.

Now, while it is not really appropriate here to go deeply into the quantitative analysis techniques which would be required for such a capability, personal system design experiences lead to the suggestion that a delivery system should be programmed (or otherwise be able) to perform tasks like the following, each of which plays a role in determining the effectiveness-efficiency nexus:

1. Recognize that potential clients will differ widely in both their socioeconomic and casualty properties and that these differences will argue that they can be placed in fundamentally different coverage modalities if the system is to make effectively optimal use of resources. Hence, a qualitative decision-support tool has to be incorporated in the design to help make *congruent* assignments of specific individuals to fundamentally different coverage modalities.

2. Allow different eligibility periods to be assigned to individuals with different socioeconomic and casualty properties. These variable eligibility periods may be used to adjust system overhead and the probability of clients' abuses of the system, etc.

3. Allow the system manager to have dynamic control over referrals and cost (and quality) of system transactions by permitting him to simultaneously monitor cost-benefit parameters of treatment providers.

4. Incorporate a real-time fiscal control mechanism through which the system manager may maintain point-in-time control over the expenditure of resources on either a time-based or capitation-based dimension.

5. Lend the ability to jointly optimize both client and system objectives through an "intelligent" referral mechanism. That is, both client dignity and benefit *and* system fiscal considerations are treated simultaneously in terms of "internal" allocations.

6. Support implications for planning as well as for day-to-day management. When properly employed, data may be gath-

ered which will enable policy decisions to be made with respect to:

    a. Optimal locations for additional service providers or agency outlets, etc.

    b. Alternative cost-benefit levels which would be associated with different budget levels or treatment policies, etc.

    c. Long-range budget and coverage package projections which can be made on the basis of accumulating historical data and experience, and combining judgmental and empirical data via Bayesean statistical* operations

7. Finally the system should, while performing operational tasks in the current period, enable policy and decision makers to "learn" gradually about the properties of their particular system, its community context and about alternatives which are available in the long run. In short, the system should provide a concatenated, empirically validated data base to support "rational" decisions and, at the same time, provide ammunition for defending against ill-conceived decisions. This is because, as the system moves into implementation, information may be collected on decisions which are approachable only by personal experience during the decision phase. Among them are:

    a. Decisions about whether mainstream or government-supported social services are fiscally preferable

    b. Decisions about which components of the system are most productive, and which are marginal or unfavorable (in cost-effectiveness terms)

    c. Decisions about what problems are encountered most frequently by clients and about the resultant cost functions of the system.

    d. Decisions about what levels of coverage or service can be provided at what budget levels, etc.

    e. Information about secular changes in the properties of the client population itself

Just why these particular system properties are desirable will not really be fully explored until the final sections of this volume. Their inclusion here is just to make the reader aware that there is an emerging technology for social service manage-

---

*For notes on Bayesean procedures, see R. Schlaiffer's *Probability and Statistics for Business Decisions* (New York: McGraw-Hill, 1959).

ment that is different than that which would be employed to manage the prototypical business firm. These properties represent the attempt to make the delivery system responsive to the technological demands that evolve in the social service sector, particularly with regard to the implications of Figure 2.1. But these technical properties cannot operate unless they are accompanied by another set of system attributes, these being evolved from what might be called the operational context in which social service operations may be conceived. In short, every system design will have to respond to a set of criteria that are local or specific in origin, as well as to the generic (enveloping) criteria set out previously. It is to these contextual criteria that we now turn for a very brief treatment.

Contextual (Local) Criteria

It is not necessary to spend much time on the matter of local or contextual criteria, for these are likely to be different for almost any delivery system design exercise. The reason is clear: these criteria are the local policy constraints imposed on the system designers, whereas every system would be expected to meet the more generalized criteria discussed in the previous sections of this chapter. At any rate, the criteria which follow are the type of specific constraints delivery system designers might be expected to encounter rather frequently:

1. The system should be a *staged* one, such that the system manager may implement any of its options, as demanded, without having to pay an overhead penalty for those options not required during a particular period.

2. The design should permit a *hybridized* system, one which may be entirely computerized or entirely manual or any combination; as the number of transactions and/or available budget increase, the extent to which computerization can be economically defended increases.

3. The system should have a capability to produce reports of a widely differing nature on demand—*on-demand report generation*. The information stored in data bases should permit complete histories to be developed on any or all of the following dimensions, singly or in combination: a. client (any

of several subdimensions), b. provider, c. payor, d. diagnosis/problem, e. treatment, f. cost, g. date.

4. The system should be *modular* in that it can accommodate, without structural alteration, any categorical program which might eventually be included in the schedule of services offered by the project the system is designed to manage.

5. The system should be *horizontally expandable* in that its logic can extend to the management of an integrated, broad-based social services delivery system (e.g., given a medical delivery system, services such as public defenders or food stamps can be added without any major modification of the system design).

6. The system should be *portable* in that it can be implemented, without modification, in other communities and at any level of government (e.g., the system could be used to manage, without essential change, a national health care delivery system or the system for a very small rural county, etc.).

7. The system should be internally *buffered,* such that errors occurring in one place do not resonate throughout the whole program; or when one system aspect must be shut down, the remaining aspects of the system are unaffected and will continue to operate with impunity.

8. The system should have some level of information *protection;* client data should be secure from outside access through a method of functional encoding, and subusers may use the systems facilities in a way which will protect their data from other subusers and also from the system manager if desirable for some reason.

It may often be the responsibility of the system designers to assist the local policy makers or program sponsors in developing the local or contextual criteria the design is to meet. For the formally constituted consulting complex, there is often a difficulty here. They would tend to consider the specifications for the system as part of the R.F.P. and as constraints to be accommodated rather than negotiated. This stems from the fact that a preferred consulting method is to treat the design problem in two phases, each with a separate set of contract procedures. First would be the *study contract*, where certain consulting operatives would be paid for developing criteria (and often for contributing to the production of the R.F.P. itself).

Second, the design function now becomes a *working contract*, and it is here that the actual design functions are performed, using the previously generated criteria as constraints. Because of the complexity of social service delivery system functions, the study and working contracts and tasks should be incorporated in a single contract. In virtually every case, the criteria set as a priori constraints for the working designers are suboptimal and sometimes set out procedures or principles which are simply not workable. This may happen, most often, because of the following: the individuals who develop the system criteria under a study contract may have little or no technical background; as is well known, the consulting corporation's "salesmen" can often be quite indiscriminate about the kind of operations to which they commit their engineering or technical personnel. Even when the salesman's enthusiasm can be contained long enough to get a technical opinion on what he is proposing, there is the simple fact mentioned earlier: in the rapidly changing, protean environment of the social service sector, yesterday's optimum is virtually always today's suboptimum. It is for this reason that we strongly recommend an admixture of the study and design functions and the employment of individuals or concerns who are less interested in terms of contract than in the substance of the system architecture exercise itself.*

It will later be shown how it is that the several local-contextual criteria discussed above were actually designed into a prototypical health care delivery system. For the next few pages and to conclude this chapter, it is well to say a few words about the procedures for managing the system design process itself and introduce the reader to the components of the HCDS we shall use as a case study.

---

*In practice, this means that we will try to force a feedback relationship between normative criteria and implementational exigencies, or between objectives and accomplishments.

BASIC DESIGN PROCEDURES

In the few pages left in this chapter, little can be done about the theory of system analysis and design. As the reader is surely aware, literally hundreds of texts have appeared over the last decade outlining the logic and mechanics of this rapidly maturing field.[3] What will be done here is to briefly discuss the type of design management scheme used with some success in several different areas and, most particularly, in the matter of the health care delivery system. By introducing not only the procedures employed there, but also reproducing the draft project logic charts and task schedules, etc., the reader will get a feel for the way in which reasonably large-scale design projects can be lent some control and coordination.

Initially, as already suggested, there is the necessity for the system designers to produce that normative system design, keeping in mind the type of criteria and constraints which were just discussed. Again, the purpose behind the normative system design is twofold:

1. It attempts to inject efficiency (or economy) into the design process by allowing the designers or analysts the opportunity to display *what they already know*. If the design team is properly constituted, they may many times find that they need collect no additional information, or will effectively minimize the expensive, time-consuming data-collection activities that characterize so many traditional system design exercises. It is only by first assessing what is already known that a reasoned determination can be made about that information yet required.

2. It allows that important attribute earlier suggested . . . the injection of originality and innovation into the design at the preliminary stages. In short, by asking analysts to design the *ideal* system, without forcing them to refer to precedents or to rely entirely on existing technologies, you are able to take best advantage of the designer's technical qualifications and also, in the most explicit way, give vent to the creative process that characterizes all good system design efforts.

In my own projects, I construct a very small, close-knit group to initiate the normative design process. In particular, a professional system theorist/analyst is chosen to head the team.

In all cases, he should have the advantage that is most particular to proper system operative: that he be facile in both qualitative and quantitative analysis. In practice, this means that he should have some engineering experience, or at least be able to use statistical and mathematical tools, but that he also have some grounding in the social sciences* and also in formal logic.[4] Such individuals are not widely available, yet they are well worth searching for even when the project is strictly technical or strictly a policy problem. One can see, immediately, that the peculiar aspects of the social service sector demand this dualistic focus, for social service programs are always located in that oft-mentioned nexus between technical demands (e.g., economic or management science criteria) and contextual (e.g., social, economic, political) constraints.

At this point I might also mention a personal prejudice developed over the years: it has been my experience that interdisciplinary teams, per se, are not an adequate substitute for the broad-based individual. That is, it is generally more promising and productive to have a single individual who himself is able to integrate the technical and social science aspects of a project than to assume that this will occur automatically simply by putting together a team of individuals drawn from different disciplines. When using this latter approach (and in some isolated cases it is indeed recommended), there is the problem that there exists no "translator" between the technicians and rhetoricians and no common conceptual vocabulary. The usual interdisciplinary team may often evolve into a struggle for the supremacy of the variables or interests of one dominant discipline or of one methodological approach. When this happens, the basic purpose of the interdisciplinary team is undermined. The system man, trained both in the quantitative and qualitative methodologies—and sensitive to both the technical and scientific components of a project—is there to guard against parochialism. Indeed, the measure of his success as a project leader is simply this: the extent to which the ultimate design emerges as a syncretic rather than parochial construct.

At any rate, with a system man on board, there are at least two other types of individuals that one might seek to constitute the initial design team. Certainly, someone should be available

*In this I mean in the substance—rhetorical, normative or hypothetical—of sociology, economics, political science, etc.

who has some experience (if the project is not completely un-precedented) with the subject matter at hand, with designing medical systems, welfare processes, etc. But it is his technical expertise rather than his field experience which is perhaps most important here, for we do not want to simply reinvent the wheel. Therefore, it is the professional pride, intellectual curiosity and creative impulse of the subject expert which most concerns us, not his degree of past association with similar endeavors. Finally, one does well to try to include an individual who is especially interested in—and well schooled in—management science. In this way, the team has a system man who is primarily responsible for the management of the design effort and especially adept at pulling together the pieces of complex puzzles (and presumably also adept at pointing out logical-operational gaps that must be filled). Bringing some necessary substantive experience is the second man on the team, the subject expert; his task, essentially, is to keep the system man and the third man, the management scientist, within the bounds of relevance so far as the particular design problem is concerned (that is, to keep them from becoming too academic, too general) and to minimize the initial need for secondary research. Finally, the presence of the management expert means that attention will be specifically focused on the role of the delivery system as a management tool and ensure that it is not obsolete in its administrative properties. Where the project is a particularly large and complex one, the design team may have to include some satellite personnel reporting to (or working primarily with) one or another of the three primary functionaries. It has been my experience that small teams can do an enormous amount of work when properly constituted and directed, and that massive hierarchical design teams usually have a way of producing designs more notable for their pedestrianism than their probity.

Of course, one does not begin a design exercise completely in the dark, even when the project is unprecedented. For social service delivery systems, we generally are able to impose a super-structure that reflects the different functions which must be attended to. Particularly, virtually every social service delivery system will have to perform (or support) at least the following functions:

1. *Access and Enrollment:* the function which actually reaches out into the community and develops a client population, or

which at least serves as the constant interface between the social service program and the potential client population. As part of the enrollment function, there is the matter of determining client eligibility and sometimes of auditing the rectitude of enrollments, etc.

2. *Assignment and Screening:* This is a very complex function (and one, oddly enough, that is accomplished poorly in most existing social service systems). Basically, it recognizes that one of the most important means by which effectiveness and efficiency can be obtained (and the interest of client dignity served) is by being strict about the classification of clients, and also recognizing that within any social service venture several different *modalities* of coverage may be dictated by different client classifications. In a properly constituted system, this area receives much attention from the designers (the HCDS assignment and screening functions will be treated at much length in a later chapter).

3. *Referral and Client Service:* This is the heart of the delivery system; it is here that we actually schedule the social casualty for remedial or therapeutic treatment, and by which we can affect intercommunication among the different social service programs that may be operating in any environment. It also has important economic implications, for the optimality with which referrals are made will in large measure determine the overall cost-effectiveness of the program itself. The basis for the referral function in a properly designed system is always algorithmic (as will be shown), and a casual or subjective referral system is usually the signature of an improperly conceived and ill-managed social service program.

4. *Client Tracking and Record Keeping:* It is here that it is ensured that the social service program is attendant to the needs of the client, per se, and is constantly alert to the *effectiveness* of the service referrals. In particular, this is where the program may display the strength of its client-professional focus and become something more than a mere transfer mechanism or maintenance conduit. More specifically, it is here that the system offers the client the benefit of its advice, consultation and decision support, at least to the extent that the casualty population requires it.

5. *Reimbursement and Internal Fiscal Operations:* This is the place to husband the resources of the system, as this is the component of the system which acts to actually affect financial transfers from the public treasury to the system's providers or to its clients. It is also here that the system manager must look for guidance as to the financial integrity of the system, and for clues as to rate of resource exhaustion, etc. This is the "technical" core of the system.

6. *Reporting and Information Inventory:* This final component of the typical social services delivery system is where to prepare the reports demanded by funding sources, trustees, etc., and also maintain the history of the program's operations. This latter task thus provides the system's managers with a planning and projection capability, and also with the major source of information about desired modifications, etc. This, obviously, is the proper rationale for a management information system, per se.

It should be noted that, although virtually all social service delivery systems will involve at least these six components (or subsystems), their particular properties or operational characteristics will respond to the set of localized (contextual) criteria peculiar to the particular project at hand. For example, as the reader will see, there is a strong correlation between the criteria reported earlier in this chapter and the way these six functions were actually designed into the HCDS system.

At any rate, to continue with these notes on design procedures, there is a very definite relationship that will be found among these six subsystems in any operational delivery system. Because the functions performed by each are relatively unique (and in some cases may even be formulated by different segments of the design team), it is useful to develop some sort of *indexing scheme.* In this way, the design tasks associated with each of the six subsystems may be neatly partitioned, but also made susceptible to simple synthesis when it comes time to actually put all the pieces together and get the system on the air. In this regard, see Figure 2.2 where the simple (linear) relationships between the six functions are illustrated and where each of the functions is assigned an index number from 1.0 to 5.0.

Again, the purpose of the index numbers assigned each of the major subsystems is simply to assist in the control of the

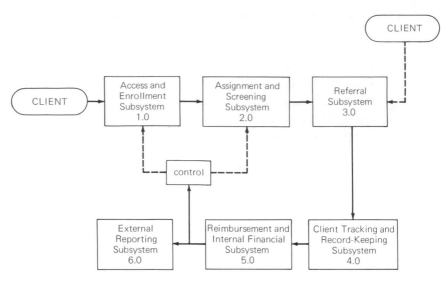

FIGURE 2.2 / Relationships between social service subsystems

system design effort itself. Given the simple relationships expected to prevail between the six subsystems (though later it will be shown that there are some complex, "organic" interconnections which emerge), it is now possible for the designers to proceed with the development of the initial project logic itself, that is, take a first cut at the actual delivery system design. In terms of general practice, the analysts might initiate this exercise by closeting themselves up and simply inventing the process by which a client might be moved through the system, relative to the several subsystems. What this involves, then, is the elaboration of each of the six subsystems and a specification of the preliminary logical connections among them.

Construction of the Normative Logic Charts

The normative system logic, again, serves as the repository for the designers' a priori wisdom, experience and intuition. In this sense, the normative system logic is sort of a grand "heuristic," a set of ordered propositions and postulates that are all, at this stage, essentially hypothetical in nature. That is, as the process of implementation and elaboration of the design are set in motion, the normative logic is susceptible to change, modification or redevelopment on the basis of emerging a posteriori (empirical,

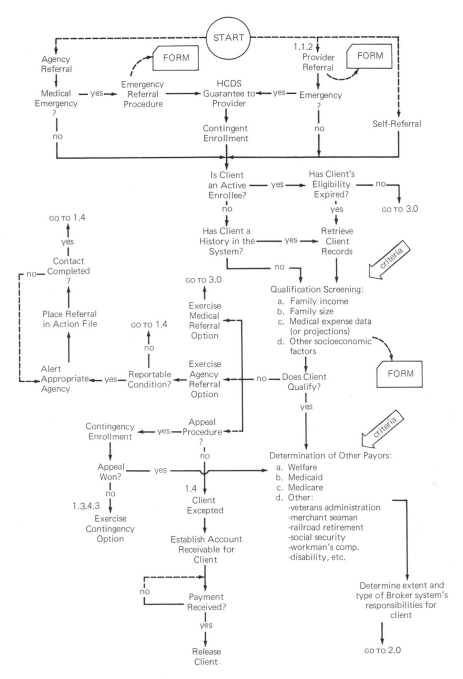

CHART 2.1 / The access and qualification processes

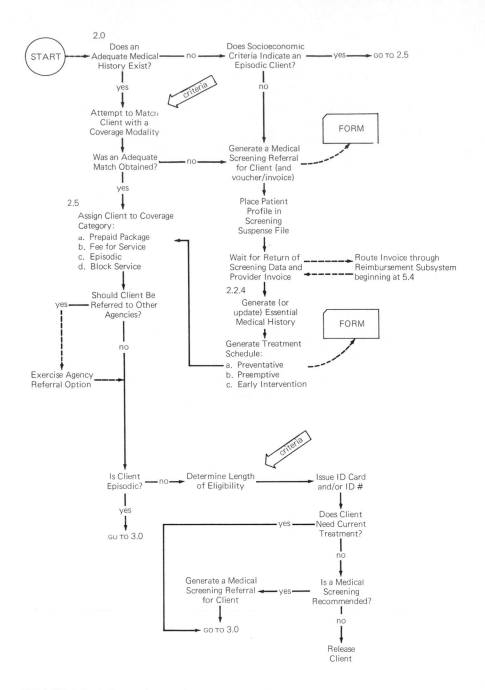

2.0

START ---> Does an Adequate Medical History Exist? --- no ---> Does Socioeconomic Criteria Indicate an Episodic Client? --- yes ---> GO TO 2.5

criteria

yes

Attempt to Match Client with a Coverage Modality

Was an Adequate Match Obtained? --- no ---> Generate a Medical Screening Referral for Client (and voucher/invoice) ---> FORM

yes

2.5

Assign Client to Coverage Category:
a. Prepaid Package
b. Fee for Service
c. Episodic
d. Block Service

Place Patient Profile in Screening Suspense File

Wait for Return of Screening Data and Provider Invoice ---> Route Invoice through Reimbursement Subsystem beginning at 5.4

2.2.4

Generate (or update) Essential Medical History ---> FORM

Should Client Be Referred to Other Agencies?

yes

no

Generate Treatment Schedule:
a. Preventative
b. Preemptive
c. Early Intervention

Exercise Agency Referral Option

Is Client Episodic? --- no ---> Determine Length of Eligibility ---> Issue ID Card and/or ID #

criteria

yes

GO TO 3.0

Does Client Need Current Treatment? --- yes

no

Generate a Medical Screening Referral for Client <--- yes --- Is a Medical Screening Recommended?

no

GO TO 3.0

Release Client

CHART 2.2 / Screening and coverage assignment logic

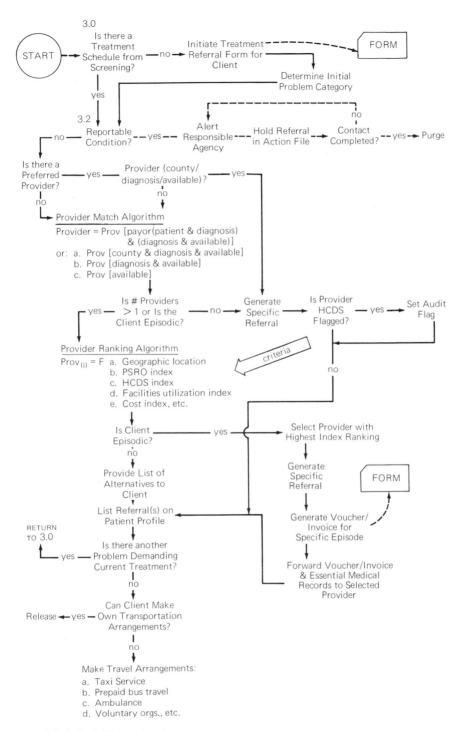

CHART 2.3 / Referral subsystem

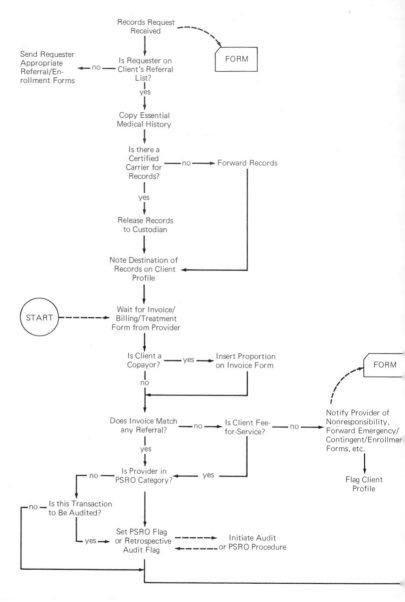

CHART 2.4 / Client-tracking and record-keeping functions

contextual) constraints and information. As a general rule, there is no particular technology of normative system design that can be imposed on all projects. But, as was just suggested, the social service system designer will at least know the several central

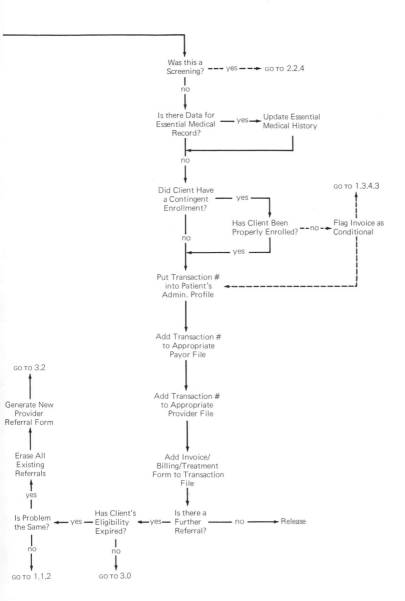

functions the system must perform and should also be prepared to construct at least three separate normative trajectories: (a) the logic of the system as viewed from the *client's* standpoint; (b) the logic that will be necessary to protect the fiscal integrity of the

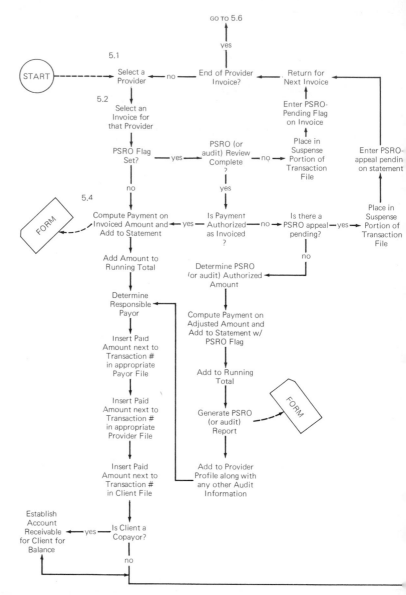

CHART 2.5 / Reimbursement, fiscal and reporting processes

system . . . the system as viewed on the *dollar dimension;* and (c) the logic needed to develop the information base necessary to affect the client and fiscal logics, and to meet reporting requirements—the *data dimension.* This means that the social service

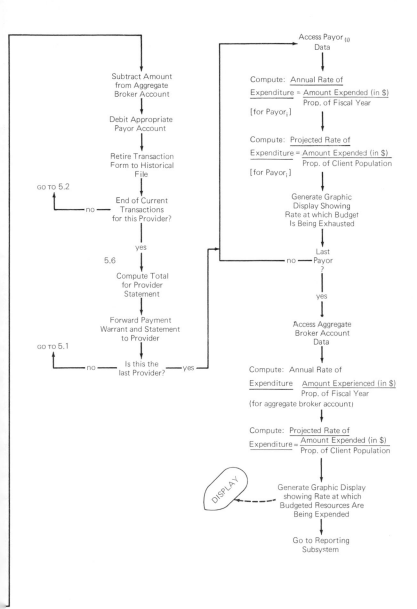

system design process—at least in the normative stage—becomes comprehensible as an exercise in successive synthesis of different viewpoints.

Now, rather than continuing to discuss the abstract aspects

of normative system design, reproduced here are the normative logic charts developed to initiate the HCDS prototype system which will later be studied in detail. These logic charts were modified, in some areas, to produce the *referent* logic charts of Chapters 4 and 5 (where the final client, fiscal and information attributes of the HCDS system are detailed). Therefore, the charts which follow here are just to give the reader an idea of what the output from a normative design exercise looks like and to give him some passing familiarity with the components (and relationships among components) of a typical social service delivery system. The reader will also note that major tasks are all given an index number—and even places where forms or files are required are indicated—so that the normative logic charts may be used to schedule the tasks which must be carried out to bring the system to life and substance. Therefore, a glance at the following constructs is useful, since much that is subsequently done here derives directly from them.

At this point, it is not important that the reader comprehend the individual functions contained in these master logic charts. Their purpose and role will all be explained in detail in the chapters which follow. What is important, again, is that this is the kind of output to be legitimately expected of a well-constituted design team, and that the master logic charts set the stage for all subsequent design activities. It is also important to understand that the production of the master logic charts is the output from a single sitting by the design team (although this single sitting may, of course, extend over several days or even weeks). The point here is for the team to be bold, but dedicated, captured by the task, as the development of a coherent normative logic for a complex project is the single most difficult aspect of the design exercise. With this illustration of what was actually accomplished in a typical design exercise (in terms of the master logic charts for the HCDS), we can leave this subject and proceed to the next step—the production of the task schedule.

The Task Schedule and Design Coordination

As just suggested, the master logic charts set the stage for all subsequent design exercises. Again, the output from the design stage is normative and preliminary, and thus serves as a guide for detailed design efforts. Nothing should be "cast in concrete" at this point. Rather, the master logic charts may be thought to set out the most promising trajectories for further inquiry and serve as the tentative outline of the design exercise as a whole. This means, in essence, that the master logic charts become the source of the specification for the substantive design tasks themselves, as they will all be directed at fleshing out or operationalizing the functions contained in the normative construct. The index numbers serve us here. Again, rather than treating the development of a task schedule in abstract, reproduced here will be the actual task schedules that were developed from the master logic charts just presented. But a few words of explanation are useful before setting them out.

Initially, the index numbers assigned the various functions and subfunctions in the master chart now become the identifiers for the specific design tasks to be performed. This is a fairly obvious tactic without need for further development. But there are two coordination functions which must be considered when developing a task schedule from the master charts. First, there is the matter of sequencing the various tasks, as the completion of some tasks will be dependent upon the prior completion of others. A recommendation in this matter is to take a sample of the functions specified in the master logic charts and rough out a simple network (perhaps following, in a casual way, the mechanics of the various PERT logics available to the modern project manager).[5] Given this simple network formulation, it should be possible to suggest approximately how many *sequence classes* will be needed. Each task will then be assigned to a particular sequence class, which will reflect either its priority to the design effort, or its rough placement along the critical path, (its positioning between tasks which are prerequisite to it, and those for which it is a prerequisite).

The second aspect of any task schedule is the assignment of specific individuals (or groups of individuals) to be responsible for the completion of the task. The algorithm for assigning individuals can sometimes be very complex. Probably the most important assignment criteria would be the following: *assign each task to that individual who needs the least additional information to complete it.*[6] The sequence of the tasks must also be considered, for the work of any individual should not be concentrated too heavily in any single sequence set.

Finally, it should be noted that the tasks which will serve to flesh out the system design imply several different types of deci-sion criteria. That is, they suggest that different types of func-tionaries will have authority over the way a task is performed or will at least have a consultative (or veto) role. Therefore, it is useful to specify, as part of the task schedule, the particular deci-sion level with which each task is associated. In virtually every social service system design project, at least three different deci-sion levels must be contemplated:

1. *Administrative Tasks:* ones that imply a decision to be made by the operational personnel of the system (e.g., by the sys-tem manager rather than by the design team)

2. *Technical Tasks:* those that involve the actual establishment or maintenance of the forms and documents required for the data processing operations of the system, or other pro-gramming or "engineering" functions (data information, al-gorithm construction)

3. *Policy Tasks:* functions which require a policy-level decision that must be made by some authority superior to the system designers or system manager (e.g., who is to be eligible for assistance, what copayor provisions to include); thus, the final system will have a front end responsive to changing policy parameters

With this very brief introduction, displayed in Table 2.1 are the actual task schedules developed for the HCDS project, with the task schedule being segmented into sections corresponding to the major subsystems outlined in the master logic charts (and with the task identification numbers being the same as the indi-vidual index numbers of the master logic charts). The decision level for each task, and the sequence into which it has been placed, are also indicated.

Finally, to give some idea of the way in which the task schedule is balanced in terms of sequencing and decision respon-sibilities, it is useful to produce separate tables (Tables 2.2 and 2.3) on each of these dimensions.

It should be noted that in the normal system design exercise, you can expect distributions of tasks pretty much like those in-dicated. That is, as for the matter of the various decision levels, you can expect the fewest number of policy decisions and the most technical decisions, etc., such that there is a frequency

TABLE 2.1 / Task Schedule

1. ACCESS AND QUALIFICATION

Task No.	Type	Task Description	Seq. No.
1.0	T	Design Access/Encounter Form (Form 2.30)	I
1.1	P	Determine Criteria for Medical Emergency	III
1.2	T	Design Contingency Enrollment Form (Form 3.40)	III
2.1	P	Determine Eligibility Criteria	IV
2.1.7	P	Determine Reportable Conditions	III
2.1.7.1	T	Design Agency Referral Form (Form 2.40)	III
2.1.8	T	Create Action File (File 2.60.10)	II
2.1.6	T	Create Agency Information Index (File 2.20.20)	II
2.3	T	Create Third-Party Payor File (File 6.30)	II
2.0	T	Design Client Profile I (Form 5.20.10)	I
2.3.1	P	Determine Conditions Broker Will Cover	I
2.3.1.1	T	Create Problem/Provider Index (File 8.40)	II
2.3.2	P	Develop Copayor Logic (Criteria/Percentage)	IV
2.2.3	T	Develop Acct. Receivable File (File 4.40.10)	V
2.2.3.1	P	Develop Acct. Receivable Logic (Copayor)	II
2.2.3.2	P	Develop Acct. Receivable Logic (Contingency)	II
2.2.4	P	Develop Definitions for HCDS Status Codes	II
2.2.4.1	T	Design Exception Notice (Form 4.80.10)	III
2.2.5	T	Create Waiting File (File 14.0)	V
2.0.1	T	Create Client Profile I (File 5.30.10.10)	I

Where P = task, T ' technical task, A = administrative decision

2. SCREENING AND COVERAGE MODALITY ASSIGNMENT

Task No.	Type	Task Description	Seq. No.
2.6.1	A	Produce Access Information Sheet (explaining criteria for socioeconomic and medical classification, HCDS eligibility criteria and procedures for proposing an individual for possible enrollment—to accompany Form 2.30 to be sent to various access agents)	I
2.6.2	A	Determine Who Should Receive Copies of Form 2.30	I
2.6.2.1	A	Determine How Many Copies to Print and Forwarding Scheduling	IV
2.5.2	T	Design Screening Referral/Voucher (Form 7.80.40.10)	III
2.5.2.1	P	Determine Who Should Screen (e.g., a block clinic or any provider)	II
2.5.3	T	Create Screening Suspense File (File 7.80.40.20)	IV
2.5.6	P	Determine Whether or Not to Implement E.M.R. (Form 5.10.30 and File 5.30.10.30)	II
2.5.6.1	A/T	If Yes, Consult with Physicians on Contents/Use	III
2.5.6.2	T	If Yes, Create File 5.30.10.30	V
2.5.6.3	T	If Yes, Design Form 5.10.30	V
3.0	A	Produce Decision Diagram for Assignment of Coverage Modalities (Form P.R. #2)	I
3.2	P	Determine Various Eligibility Periods (given modalities)	II
3.3.1	P	Determine Contract Terms for System Enrollment	III
3.3.2	T	Design ID/Contract Forms	IV

Table 2.1 *cont.*

2. SCREENING AND COVERAGE MODALITY ASSIGNMENT

Task No.	Type	Task Description	Seq. No.
3.3.3	T/A	Determine Algorithm for Translating Social Security Number into Client Identification	II
3.3.4	T/A	Determine Payor Identification Codes to Prefix Client Identification	IV

3. REFERRAL LOGIC

Task No.	Type	Task Description	Seq. No.
4.0.1	T	Design Referral Form (Form 8.30)	II
4.0.2	P	Determine if Form 8.30 Should Be Distributed to Agents in the Field or Held Internally	I
4.3.2	T	Design Referral Exception Notice (Form 14.0)	IV
4.3.3	T	Create Exception File (File 20.0)	V
4.4.2	P	Determine Criteria for Acceptable Providers	III
4.4.2.1	T	Develop Index of Authorized Providers as Header to File 8.40 (by specialty)	IV
4.5.1	T	Write Program to Translate CHPA Provider Data to HCDS Format for File 1.20.20 (physicians)/Sort	I
4.5.2	T	Create Temporary Inpatient Provider Files (1.20.30–1.20.50.20) until CHPA Data Available	III
4.5.3	A	Assign Initial HCDS Audit Index Values to Authorized Providers	V
4.5.4	T	Develop Algorithm for Using Client Input (problem, coverage modality, Third-party coverage, geographic area, etc.) to Generate a Set of "Best" Provider Alternatives	III
4.6.1	A	Determine Number (n) of Alternatives to Provider Clients	II
4.6.8	A	Determine Voucher-Handling Logic for Block Clients (e.g., should voucher be given to client or sent to clinic)	III
4.7	A	Establish Procedures for PSRO Concurrent Audit (when available) for Ambulatory Transactions	IV
4.8	A	Determine Available Transportation Modalities (prepare decision criteria such as those prepared for coverage modalities in PR #2)	V
4.6.3	T	Create Referral Archive File (File 8.30)	V
4.5.4.1	T	Develop Procedures for Updating Provider File(s)	III
4.5.4.2	T	Develop Procedures for Producing Weekly (e.g.) Copies for Manual Referrals	II
4.5.4.3	P	Determine if Referrals Should Be Made in the Field (by e.g., public nurses) or Decentralized?	I

4. RECORD KEEPING AND CLIENT TRACKING

Task No.	Type	Task Description	Seq. No.
5.1.3	T	Design Provider Invoice (Form 12.10.30) (in consultation with Physicians)	I

Table 2.1 *cont.*

4. RECORD KEEPING AND CLIENT TRACKING

Task No.	Type	Task Description	Seq. No.
5.1.3.1	T	Create Provider File II (Form 5.10.20; 5.30.10.20)	II
5.1.3.2	T	Determine How Many Forms to Print and Distribute Algorithm	IV
5.1.6	T	Create Holding File (File 12.80.10) and Segment: 1 = PSRO Audit; 2 = HCDS Audit; 3 = Contingency; 4 = Suspension	IV
5.3	T/A	Design Notice of Nonresponsibility (Form 12.50)	III
5.2.3	T/A	Establish General Account (File 6.50.20.50)	II
5.3.5	T	Design Suspension Notice (Form 16.0)	IV
5.6.1	A	Determine Internal Audit Procedures (it is recommended that samples of invoices be taken and examined for double charging by providers against previous invoices for any client/diagnosis/date combination)	III
5.7.2	T	Create Transaction Summary File (File 12.10.10)	II
5.7.4	T	Create Current Transaction File (File 12.10)	II

5. REIMBURSEMENT AND FISCAL

Task No.	Type	Task Description	Seq. No.
6.3	A	*Note:* The Provision for Examining Invoices for Overtreatment, Undertreatment, Overcharging or Undercharging Cannot Be Implemented until there Is an Adequate Data Base. Therefore, Invoice Data as to Treatment/Diagnosis/Cost Correlations should be Recorded on a Separate File from Which, after the First Year's Operations, for example, Probability Distributions of the Type Discussed in Chapter 4 Can Be Generated. These Distributions Should Be Able to Be Created from the First Year's Transaction Summaries (from File 12.10.10)	III
6.5.1	T	Design Provider Statement (Form 15.0)	IV
6.5.4.1	T	Create Payor Files (6.50.20.10–6.50.20.50)	V
6.5.4.2	T	Create Payor Working Statement (for aggregate debit for a particular reimbursement period—throw-away sheet)	V
7.1	T	Develop Algorithm for Computing Annual Rates of Expenditure (time or capitation budgets)	II
7.2	A/T	Create Formats for Fiscal Displays	III
6.3.1	T	Select Diagnosis/Treatment Code Source (if immediate implementation of quality/cost control)	I
6.3.1.1	T	Select N Most Frequently Expected Diagnosis	II
6.3.1.2	T	Associate Treatments with Diagnosis	III

Table 2.1 *cont.*

5. REIMBURSEMENT AND FISCAL

Task No.	Type	Task Description	Seq. No.
6.3.1.3	T	Associate Cost Standards with Treatments	IV
6.3.1.4	T	Develop Values of HCDS Audit Reductions (per incident)	V

6. ANCILLARY TASKS (not specifically related to any single subsystem)

Task No.	Type	Task Description	Seq. No.
0.1.0	T	Prepare Translation Tables	IV
0.1.1	T	Ta: Diagnosis Codes → English	IV
0.1.2	T	Tb: Treatment Codes → English	V
0.1.3	T	Tc: Provider Identification Codes to Address Labels	V
0.1.4	T	Td: Client Identification Codes to Address Labels	V
0.1.5	T	Te: Access Agent Identification Codes to Address Labels	V
0.2.0	T	Develop Procedures for Updating Files: – 1.20.20 – 6.30 } Indices – 8.40 – 2.20.20	III
0.3.0	A	Determine Nature of Block Contracts and Select Providers	III
0.3.1	A	Determine Which Rates and Numbers Should Apply to Each	IV
0.3.2	A	Determine Feasibility of "Funded" Insurance Coverage, Select Carriers for Quotes, etc.	II
0.4.0	A	Prepare Information Brochures for: Clients—Access Agents—Providers	IV
0.4.1	A	Develop EMR (Records) Request Logic (Form 11.10)	V
0.4.2	A	Develop Ombudsman Logic (Form 14.50)	V
0.5.0	A	Develop Appeals Logic (for Clients)	IV
0.6.0	A	Develop PSRO Interface and Determine their Reporting Formats and Procedures	III
0.7.0	T	Develop RDG Front End for Generating On-Demand Reports from Transaction Summaries (File 12.10.10)	III
0.7.1	T	Collect Reporting Requirements for Payors (General)	II
0.7.2	A/T	Determine Audit Requirements of County for the General Account (File 6.50.20.50); Other Payor Requests for an Audit Trail	III
0.8.0	A/T	Produce Narrative for Project Package	I

distribution that would be crudely *skewed to the right.* By the same token, you can expect a *roughly normal* distribution of tasks by sequence, such that the majority of the tasks tend to be completed in the middle portions of the project continuum. And there is one final thing to mention: in actual practice, each of the sequence sets would be bounded by actual dates. That is, the

TABLE 2.2 / Distribution of Tasks by Area of Responsibility

Policy	Administrative		Technical		
1.1	2.6.1	5.6.1	1.0	4.0.1	5.7.2
2.1	2.6.2	6.3	1.2	4.3.2	5.7.4
2.1.7	2.6.2.1	7.2*	2.1.7.1	4.3.3	6.5.1
2.3.1	3.0	0.3.0	2.1.8	4.4.2.1	6.5.4.1
2.3.2	3.3.3*	0.3.1	2.1.6	4.5.1	6.5.4.2
2.2.3.1	3.3.4*	0.3.2	2.3	4.5.2	7.1
2.2.3.2	4.5.3	0.4.0	2.0	4.5.4	7.2*
2.2.4	4.6.1	0.4.1	2.3.1.1	4.6.3	0.1.0
2.5.2.1	4.6.8	0.4.2	2.2.3	4.5.4.1	0.1.1
2.5.6	4.7	0.5.0	2.2.4.1	4.5.4.2	0.1.2
3.2	4.8	0.6.0	2.2.5	5.1.3	0.1.3
3.3.1	5.3*	0.7.2	2.0.1	5.1.3.1	0.1.4
4.0.2	5.2.3*	0.8.0*	2.5.2	5.1.3.2	0.1.5
4.4.2			2.5.3	5.1.6	0.2.0
4.5.4.3			3.3.2	5.3*	0.7.0
			3.3.3*	5.2.3*	0.7.1
			3.3.4*	5.3.5	0.7.2*
					0.8.0*

*Indicates a joint responsibility

TABLE 2.3 / Distribution of Task by Sequence in Time

I	II		III		IV		V
1.0	2.1.8	5.2.3	1.1	4.6.8	2.1	0.1.0	2.3.3
2.0	2.1.6	5.7.2	1.2	4.5.4.1	2.3.2	0.1.1	2.2.5
2.3.1	2.3	5.7.4	2.1.7	5.3	2.6.2.1	0.3.1	2.5.6.2
2.0.1	2.3.1.1	7.1	2.1.7.1	5.6.1	2.5.3	0.4.0	2.5.6.3
2.6.1	2.2.3.1	6.3.1.1	2.2.4.1	6.3	3.3.2	0.5.0	4.3.3
2.6.2	2.2.3.2	0.3.2	2.5.2	7.2	3.3.4		4.5.3
3.0	2.2.4	0.7.1	2.5.6.1	6.3.1.2	4.3.2		4.8
4.0.2	2.5.2.1		3.3.1	0.2.0	4.4.2.1		4.6.3
4.5.1	2.5.6		4.4.2	0.3.0	4.7		6.5.4.1
4.5.4.3	3.2		4.5.2	0.6.0	5.1.3.2		6.5.4.2
5.1.3	3.3.3		4.5.4	0.7.0	5.1.6		6.3.1.4
6.3.1	4.0.1			0.7.2	5.3.5		0.1.2
0.8.0	4.6.1				6.5.1		0.1.3
	4.5.4.2				6.3.1.3		0.1.4
	5.1.3.1						0.1.5
							0.4.1
							0.4.2

sequence sets refer to time intervals on a continuum from project initiation (at time t_0) to projected completion (time t_n), much as follows:

	I	II	III	IV	V	
$(\emptyset)$	(t_1)	(t_2)	(t_3)	(t_4)	(t_5)	(n)

The interpretation is the obvious one: the t_1, t_2, etc., represent the date at which the finished output for each task sequence is expected. That is, each of the tasks associated with say, Sequence III, are to have been completed by (t_3). For this reason, it is important to distribute tasks for individuals more or less evenly across all intervals, and to keep a careful check on the productivity of each of the project functionaries with respect to their own work sequences.

Of course, there will likely be interruptions in the neat project schedules outlined above. Adjustments will have to be made, slippages recorded and perhaps even some personnel replaced. But it is important to have a neat, simple schedule so that remedial action may be taken promptly and with some probability of success. For very large, well-financed design projects, the project leader may have a rather elaborate PERT chart or other network device at his personal disposal. Even so, it is a good idea to have a very simple sequence chart and task list available for those personnel who find PERT charts and formal network models somewhat awesome.

This is about all that can be said at this point about the mechanics of system design. As suggested, this has been a discussion mainly of very mundane, practical matters; those searching for elegant discussions of technique and the various system design algorithms can go to one of the many fine texts in the area. As for us, here, we shall now move on to analyze the structure and operating logic of a typical social service delivery system and, in the process, amplify many of the arguments introduced in this chapter.

NOTES AND REFERENCES

[1] The analysis presented here of the consulting complex is deliberately critical and would refer only to a minority of consulting firms. A much kinder appraisal of consulting operations—but a useful reference—is that given by Garry D. Brewer in *Politicians, Bureaucrats and the Consultant* (New York: Basic Books, 1973).

[2] The technical points made here are considerably expanded and defended in Chapter 3 of John W. Sutherland, *Administrative Decision Making* (New York: Van Nostrand Reinhold, 1977).

[3] For a survey of the design technology (and for a list of other references to which the reader might turn), see John W. Sutherland, *Systems:*

Analysis, Administration and Architecture (New York: Van Nostrand Reinhold, 1975).

4 One instrument that is particularly important for conducting qualitative or normative exercises is the Delphi technique. For more on this, see Linstone and Turoff eds., *The Delphi Method: Techniques and Applications* (Reading, Mass.: Addison-Wesley, 1975).

5 For an introduction to the construction of PERT networks, see McLaren & Buesnel's *Network Analysis in Project Management* (London: Cassell & Co., 1969) or D.C. Robertson's *Project Planning and Control: Simplified Critical Path Analysis* (Cleveland: CRC Press, 1967).

6 This point is derived from an argument given by Adrian McDonough in *Information Economics and Management Systems* (New York: McGraw-Hill, 1963).

3

ANATOMY OF A DELIVERY SYSTEM

INTRODUCTION / The purpose of this chapter is to take a broad overview of a social service delivery system and to address matters of both structure and process. Moreover, it will try to suggest something of the limitations of modern management technology, for we simply do not have all the tools needed to completely discipline the administration of these complex organizations. Therefore, the task here is to reflect on a feasible managerial technology, one that suggests the best that can be done given the relative immaturity of social service technology.

In terms of the scope of the discussions, this chapter will look at a typical social service system which entails the functions which virtually all social service delivery systems must be prepared to perform. Therefore, the arguments and prescriptions will be rather generalized, and not specific to any single system. This will contrast with the specific arguments of the subsequent chapters of this volume which detail the faculties built into a real-world system.

THE LIMITS OF MANAGERIAL TECHNOLOGY

In no case will the design of an effective-efficient social service delivery system be a straightforward matter. This again brings up the matter of managerial integrity. There is simply no way that the system designer can build into the system model *all* the techniques or programs required for rational system management. That is, no matter how clever the design, some technical and analytical capabilities will be presumed of the system manager and his staff, the system implementors. Again, properly qualified system managers will be scarce indeed, perhaps even more scarce than properly qualified system designers. Therefore, the feasibility of a system design may often be limited by the lack of technical qualifications among those who are to implement and administer the system. The inheritors of the system design—the program's administrative staff—often introduce the most concrete and crippling constraints on the extent to which a normative system design may be made operational.

Some readers may ask how this could be possible, given the staggering number of educational programs, (the legion of business schools, schools of public administration and system curricula) which pretend to train modern managers. Even a casual study of these programs will answer for the paradox. The vast majority of these programs will suffer from two crippling defects, defects which carry over directly into the operational world:

1. In most large business and public administration schools, there is really no attempt made to seriously and adequately expose the student to *both quantitative and qualitative* analysis. A student may often elect to pursue one or another of these analytical tracks, but is seldom forced to sample from each equally (or sufficiently). The result is that most students entering the world of social service management (or management in general) will tend to be either mathematicians or rhetoricians with all the operational inhibitions these concentrations imply.

2. Even where there is some attempt to expose the student to both quantitative and qualitative analytical techniques, there is often a distinct *lack of instruction in the substantive bases* of administrative science. That is, the student pursues his studies of management theory and technology as if this were a sufficient preparation for assuming policy and decision responsibilities in the real world. In more specific terms, this means that many educational programs have tended to treat management as if it were independent of the content of the substantive disciplines: economics, political science, sociology, social psychology, anthropology, etc.

The more enlightened view of management studies is that they should exist at a constantly elaborating nexus between the technical and substantive (soft) disciplines, between analytics and praxis. That is, instruction in the managerial functions—marketing, finance, production, etc.*—is not a sufficient education base for modern managers, except where the manager's ambitions are restricted to the maintenance of a simple commercial enterprise: one with a single product line, a stable source of materials supply, a stationary market profile and demand posture and one operating in an environment that puts little premium on innovation.

In so short a space it is difficult to suggest why this emphasis on the managerial functions is misplaced, but an effort will be made here. Initially, when looking at the generic functions of management—marketing, production, finance, etc.—we see that the way in which they must be performed will be dictated by the context in which the organization operates. In some contexts marketing will be a simple, mechanical exercise; in others, it will be confounded by the most intractable and complex factors.[1] Therefore, there should ideally be a *unique* marketing technology for each unique milieu (or operational environment) we might define. As taught, marketing technology tends to be a monolithic structure, a set of concepts that are either so abstract that they relate to no particular context, or a set of recipelike procedures that may be functional only in the most simple and placid environments. The same tends to be true of the instruction given in the other functional domains of production and finance, etc. It happens that instruction of the first type (the abstract perspective) tends usually to be rooted strongly in quantitative tools, while the instruction that pertains to the simplest milieu—the recipe-based, pragmatic programs—tends to be largely rhetorical in nature (e.g., treating the managerial functions as distillations of actual experience, and thereby bereft of theoretical or generic significance).

The tendency is to explain away these different emphases in the following way: The *better* schools—catering to brighter, more ambitious students—will stress the theoretical, intellect-expanding curriculum, while the junior colleges and "diploma mills" will stress the pragmatic emphasis. Moreover, some might say, it is only proper that business schools and other faculties should attempt to differentiate their product, catering to the different interests of different students. But neither of these de-

*Public administration curricula often merely generalize the emphases.

fenses is quite true. First, the "better" schools are not always those with the strongest theoretical base; some are blatantly pragmatic and instrumental. They survive, and their reputation maintains itself, not because of the pioneering quality of their course offerings, but because they have connections to the world outside and can obtain prestigious placements for many of their graduates. Among such schools there will be a definite strategy (though often tacit) to make sure that the graduates reflect the properties of existing managers with whom the school maintains relations. That is, these schools may think it inadvisable to impose graduates who are significantly better trained than are the members of the existing managerial corps. Therefore, these schools tend to dampen managerial innovation, and therefore pose no threat to the traditional managerial establishment. More seriously perhaps, the relations between these "realistic" schools and the business and government community are often dictated by a kind of implicit quid pro quo arrangement: the university faculties rely on the business and government organizations to provide research support and other financial emoluments and to provide places for their graduates; the prevailing business and government structure, on the other hand, relies on the schools of business administration and public administration to be moderate and reticent in their criticism. It is thus that some very powerful and prestigious business and public administration faculties become servants of the very subjects they are supposed to study and criticize. It is, indeed, no accident that the most vocal and constant criticism of business and governmental practices comes from somewhere other than schools of business and government.

Be this as it may, we must answer briefly for the concept that it is good to have a wide selection of educational emphases from which prospective management students may select. From a personal perspective (which is shared with an increasing number of others), this is one of the most damaging of the assumptions enveloping modern management education. The reason is one mentioned earlier: a person simply cannot manage a complex organization if equipped solely with quantitative instruments and an abstract appreciation of organizational functions, nor can a person do an adequate job if equipped solely with an introduction to a set of pragmatic cases and familiar only with rhetorical arguments about management theory (the so-called principles of management, which are not really principles so much as they are predilections).* Rather, adequate management of the complex organizations we have evolved—be they commercial or public—demands a more or less equal facility in dealing with numbers and dealing with concepts, and a more or less equal balance between generalized theoretical bases and the applied (pragmatic) aspects of administration. In short, the pure mathematician is as much a detriment to organizational efficacy as is the abject rhetorician. And the pragmatist, eschewing any theoretical substance, is as dangerously deluded as is the generalist who refuses to involve himself at all with the vagaries of any particular context or who denies local criteria any authority.

A look at the programs from which the vast majority of modern managers have been graduated, shows such parochialism time and time again. Those who graduate from essentially quantitative curricula—the operations researchers and industrial engineers, etc.—tend to have virtually no facility for incorporating contextual or substantive variables into their work, for the analytical instruments to which they were exposed demand that all variables be empirically accessible, measurable and manipulable.[2] The social, political, economic and behavioral forces which occur constantly in the real world seldom meet any of these criteria. As such, the "technician" spends much of his energy—and thereby loses much of his real significance—in trying to make the world simpler than it really is. By the same token, the graduates of the pragmatic programs—the managers whose education was confined to carefully circumscribed sets of cases or to the mechanics of marketing, production, etc.—do not have any of the analytical skills necessary for properly simplifying the world. They, therefore, tend to constantly discount the capacity of science and modern mathetico-statistical methods to support rational management, and thus tend to be more reactive than reflective (in short, their world is a concatenation of surprises and crises).

Now, the majority of the world's problems tend to fall rather neatly through this vast no-man's land between the technician and the rhetorician. The solution of most problems, or at least those of any significance, demands a simultaneous attack by

---

*As was earlier mentioned, these are among the reasons why a traditional education in business administration does not qualify the individual to step immediately into the social service sector. As the Table of Instruments shortly to be introduced makes clear, public administration is *more* than business administration in virtually all contexts.

mathematical and conceptual methods . . . demands a balance of both discipline and imagination, both containment and humility. Thus, the truth of complex subjects is just as likely to elude the pure conceptualist as to elude the abject pragmatist. In the most specific terms then, it is a suggestion that a majority of administrative errors wish themselves on our world largely because of the fact that most significant problems will not yield to technical or rhetorical solutions alone and that the vast majority of existing managers are probably analytically and intellectually unprepared for the positions they hold. That is, to the extent that a manager is a product of a parochial education—either exclusively technical or rhetorical, exclusively theoretical or pragmatic—he must expect to encounter many problems with which he is forced to deal inadequately (even though he will probably not always be aware of his limitations).

Now these arguments are not really tangential to the purposes of this volume. It is important to understand that many of the administrative horrors in both the public and private sectors stem not just from faults of character (from the existence of venal, corrupt or manipulative managers), but also from faults of mind and intellect, which are perhaps far more pervasive and telling. But even more to the point for our purposes, the reader should not think that there really is an operational, *sufficient* technology for the management of the social services sector. In large measure the traditional parochializing influences in the management science domain has prevented the development of a well-articulated technology capable of dealing with significant complexity. Therefore, even if a curriculum were established that was determined to provide the best possible education for its students—structured to get that recommended mix between theory and practice, mathematics and rhetoric, technical and social skills—the graduate would still find that his analytical arsenal, his repertoire of techniques, would have important gaps. He would certainly be better prepared than his parochial counterpart, but he would still not be completely qualified. This is because we, as academics, have not yet developed management technology to the point where it will encompass the full range of problems the administrator might encounter. By and large, we can show the student how to solve the simpler mechanical (or deterministic) problems that abound in virtually every organization, but we can only give him incomplete guidance when it comes to matters of grand strategy, goal setting and higher-order problems impinging on the larger commercial or public enterprise.

Personal experience in auditing existing management education programs (for purposes of certification or formal review) and in managing projects designed to deal with very complex issues exposed me to the grand claims made by some operations researchers, management scientists and system analysts about current technological capabilities. The problem is this: when an individual or set of scientists has had considerable success in solving simple problems, they tend to project their capability a bit too far; that is, they tend to get a bit overconfident about their technology. It is such scientists—coupled with generally gullible and unschooled political functionaries—who suggest that because we were successful in getting to the moon or developing color television, that all problems are susceptible to our technology . . . that all we need to solve social and economic problems is an Apollo mission type attack coupled with a flow of federal funding. As any serious system scientist will say, this is simply not the case. Sending a rocket to the moon and building a color television set are problems of an entirely different order of complexity than social and economic development, than curing cancer or comprehending the genetic code. The management technology that builds rockets and television sets is simply not adequate to the more subtle and protean problems which are distributed throughout the troubled world.

But what more is needed? There is probably no single individual anywhere who can answer this question completely or unequivocally. Introduced here is a short list of areas that need more development. It is perhaps a *minimal* schedule of technological shortcomings, but it does list the technological innovations which must be generated in order to rationally administer complex organizations in general and social service systems in particular:

1. A project management technology that will lend itself to nonmechanical, indeterminate problems . . . to problem-solving exercises where, for example, the appropriate objective or goal is not fully specified and where the project's output (unlike that of the Apollo mission) is not directly measurable or tangible in empirical terms, and where the conditions of success are variable (nonstationary). In short,

we need to articulate, specify and equip modern managers with a *heuristic management technology.*[3]

2. A comprehension—and models—of *alternative organizational structures*[4] since the hierarchical, bureaucratic organizational modality is not really suitable to nonmechanical missions. Specifically, we must analyze and develop the theory of the *matrix* organization and its more complex successor, the *reticular* modality (where traditional lines of authority and responsibility become replaced by sets of "adhocratic," plastic networks).

3. A set of human relations techniques—and comprehensions—that are effective in *motivating and controlling professional personnel*[5] (engineers, scientists, etc.) on whom operating organizations are placing increased reliance. As things stand today, the majority of the "tricks" offered by the human relations experts are directed toward the manipulation of low-level personnel and functionaries within traditional bureaucratic frameworks, and toward making minor organizational adjustments that promise increased mechanical productivity. Managing the professional employee is a different thing altogether, and one we know little about. He is a complex phenomena with a potentially different set of sociobehavioral referents than his blue-collar or white-collar counterparts and a very much more complex set of motivational predicates.

4. An entire concept of management of complex systems as an exercise in *action-research,*[6] which simply suggests that, where environmental and operational conditions are protean and rapidly changing, the only appropriate managerial strategy is one which combines doing with learning. That is, for organizations with a complex mission—and dependent on innovation—a management strategy must be devised which makes concatenate education the primary objective and relegates traditional maximization-minimization criteria to a subordinate role. Moreover, it may be necessary to develop a set of management procedures that suggest how one may minimize the possibility of irrevocable dysfunctions when the probability of decision error is significantly high. Such a set of procedures would ostensibly be predicated more on the exploitation of formal logics than on the

traditional microeconomic base which underlies most existing management science concepts and techniques.

5. The technological prerequisites required for the operational *synthesis* of technical and social science criteria and for the increasingly essential interchange between the "hard" and "soft" sciences, and between engineering and normative philosophy. This implies much more than the interdisciplinary enthusiasm which signals current efforts. For a synergistic synthesis demands more than simply placing a physicist and sociologist in confrontation, or in developing a team where representatives of different interests and disciplinary backgrounds are imposed on each other.

6. *Techniques*—algorithmic processes—which must be refined or developed to do an adequate job managing complex organizations. Among the most obvious technical innovations demanded are these:

   a. A capability for computing the distributed costs of processes, e.g., contextual, social or environmental costs. Moreover, once this basic computational procedure is developed (which probably means the isolation of complex quantitative surrogates), we must develop algorithms for *internalizing* exogenous costs.

   b. A capability for developing meaningful cost-effectiveness and cost-benefit indices for the nonprofit sector, where second- and third-order effects, etc., must be accounted for.

   c. Properly articulated techniques for determining expected costs of decision error and other performance criteria that are proper conjunctions of subjective and objective probabilities . . . syncretic formulations of judgment and observation, speculation and experiment. This demands looking beyond the currently popular Bayesian statistical technology and developing dynamic programming (decision algorithms) that are not so restricted as to the number (and quality) of state-variables they can encompass.

   d. A movement toward the *reinstatement of optimality* as a respectable criteria for management performance (despite the fact that criteria like sufficiency and adequacy are popularly posed as the best that real-world managers can be expected to do). Operationalizing optimality requires the development of techniques for containing multidimensional, quasi-variables, for dealing efficiently with nonlinear stochastic systems, for enjoining qualitative and numerical factors, for modelling complex systems using topological forms and the constructs available from modern algebra, etc. It involves making better use of computers, exploiting their potential for manipulating something other than numbers; the deep commitment of modern mathematicians to the demands of policy formulation and decision making, etc.; and it involves the strict adherence of working social scientists to a system-based epistemology.

Now, it is true that some research is being done in all these technical areas and that some real progress has been made in the last several decades. But the point is this: until we have gone much, much further in all these areas, we cannot pretend to be able to train managers that are fully capable of dealing with complex problems. Even more pointedly, those business and public administration programs that exclude any of these subjects from their curriculum cannot really pretend to be turning out even adequate management personnel. Most readers will agree that, to the extent that a significant majority of those acting currently as professors of business or public administration are largely unschooled in the subjects and innovations just listed, the prognosis for the immediate future of the management sciences is perhaps not very good. In short, it will be some time before the capabilities of instructional faculties, and therefore the capabilities of real-world managers, are made consonant with the demands that the complex systems impose on us.

But, despite the fact that a genuine and sufficient management capability rests somewhere beyond the immediate horizon, no opportunity should be lost to employ the minimal capabilities and technology that are currently available. This is mentioned here because, as the outline of the logic and substance of the prototypical social services delivery system emerges in the remaining pages, it it not meant for the reader to think that we are cracking any new technological barriers or converging on a new optimality. Far from it. The hope is that what will be done here does represent an improvement—in some areas at least—over practices which currently prevail in the social services sector. In no case should the arguments outlined below be considered the end of the road. They are a good compromise between what current technology will support, and what the current corps of operational social service managers will tolerate or be able to implement. Therefore, as we take the next steps into social service technology, the reader must be aware of the distance yet to travel before we can legitimately pretend to meet the expectations society has for us.

However, despite the technological limitations and procedural lacunae just mentioned, there are ample opportunities for injecting rationality into social service administration. The social service sector may be more receptive to restoring optimality to its rightful place as a management criteria than are most business organizations, for the social service program is usually more visible than its commercial counterpart and cannot hide behind reflective performance criteria like share of market or profit rates compared to those of competitors. To the extent that social service managers are accountable at all, it is in an absolute rather than relative sense. Secondly, there are active constraints against competition in oligopolistic industries, largely because block investors (banks, insurance companies, etc.) want to discourage the possibility of any of their investments being driven out of business. So it is a naive fellow indeed who suggests, blanketly, that what government service needs is more businessmen. Just why this is the case can best be shown with reference to Table 3.1, where is set out a schedule of instruments that provide the minimal intellectual and analytical tools the rational social service manager must have in his repertoire, and which will be the foci of much of the discussion in the remainder of this volume. Also note that many of these instruments are also going to be required of business managers as the context in which they operate becomes complicated by unstable interest rates, interrupted supply sources, enforced social and political constraints, etc. Thus, in a sense, Table 3.1 is a synoptic view not only of social service management technology, but of management technology wherever considerable complexity is encountered.

Table 3.1, then, is a brief but ordered overview of the analytical instruments that should be in every social service manager's repertoire. The pages that follow will show why these instruments are useful and give some suggestions about the procedures and conditions which recommend their use. This begins with a look at the structure of the social service sector in general where

TABLE 3.1 / An Ordering of the Instruments and Procedures Constituting a Technology for Social Service Management

I. THE ACCESS SUBSYSTEM

Instrument	Formulation	Implication
1. Casualty Correlation	Determines which casualty conditions tend to appear in statistical concert with which others, relative to some geographic domain or population stratum.	a. Suggests which agencies or programs should be "linked" with which others through the *transfer* mechanism. b. The strength of the correlations would reflect the expected frequency with which different agencies will be expected to "share" clients. c. Serves to determine the configuration for an integrated social service system by rationalizing the *network* of connections.
2. Client Density Distribution Function	Shows the way in which various casualties are distributed across a domain, isolating the following patterns: 1. Randomization 2. Stratification 3. Clusteration	a. Provides the basic input to rationalize the aggregate access configuration (via the centrix model below). b. Serves as an input to the client population projection and demand functions (below) c. Disciplines the attempt to develop a rationalized "mix" of access modalities.
3. Centrix Model	For any given density distribution function, the centrix model will return an effectively optimal location for an access or service station.	Provides the system manager with a technique for locating an access or service station such that the demands on the average client (within some domain or cluster) are effectively minimized.
4. Simulation and Queuing Analyses	Will, when used in concert, generate an effectively optimal *set* (or structure) of access locations, in terms of absolute effectiveness.	The queuing analyses may be used to find a particular access configuration that minimizes client waiting time or average delay, whereas the simulation models developed from the queuing functions can suggest whether a single access location or

Table 3.1 *cont.*

I. THE ACCESS SUBSYSTEM

Instrument	Formulation	Implication
		multiple stations are most effective. The centrix model provides the indication of what locations should be subjected to simulation.
5. Simoptimization Modelling	Will yield an effectively optimal access configuration in terms of aggregate cost-effectiveness, by incorporating cost data and overhead implications into the various simulated alternatives.	*a.* Develops an operational cost-effectiveness index for the various configurational alternatives, for purposes of selection. *b.* Converges on a most favorable and flexible tradeoff between effectiveness (as a function of number of locations and level of access overhead) and efficiency (which implies minimization of locations and access overhead). The simoptimization solution is flexible in that it can be adjusted to meet changing policy parameters and client characteristics.
6. Access Elasticity Function	Measures the expected proportion of the casualty population attracted by various access modalities supported at different levels.	*a.* At the strategic level, isolates the most favorable "mix" of access modalities (1. Promotion, 2. Recruitment, 3. Search, 4. Enlistment, 5. Transfer). *b.* Suggests the resources that should be devoted to each modality (a tactical decision).

II. QUALIFICATION SUBSYSTEM

Instrument	Formulation	Implication
1. Threshold Function (elasticity)	Suggests the relationship between varying qualification criteria and client population.	*a.* Shows how the system manager may adjust client population by raising or lowering (easing or tightening) conditions for

Table 3.1 *cont.*

II. QUALIFICATION SUBSYSTEM

Instrument	Formulation	Implication
		enrollment and system coverage. *b.* In conjunction with access elasticity function, becomes an input into the demand projection function (below).
2. Effective Demand Schedule	Suggests the way in which demand for system resources varies with respect to changes in client enrollment and coverage criteria (service levels).	*a.* Used to maintain a symmetrical level of service over some time interval (or across some fixed population base). *b.* Used to estimate resource requirements relative to population-treatment projections. *c.* Permits the system manager to predict actual as opposed to desired per capita resource levels.
3. Production Function	Shows how level of system service (breadth and intensity of coverage) varies with different resource levels and client populations . . . senses the terms of the tradeoff between enrollment, service and resource levels.	*a.* Allows the system manager to control rates of resource expenditure with respect to capitated or temporal budget limits. *b.* Can be used to rationalize the tradeoff between quantity of clients treated, and quality (intensity or breadth) of service. *c.* Provides a basis for rationalizing the distribution of incremental increases or decreases in the resource (funding) bases, relative to marginal utility criteria.
4. Audit Scheduling Functions	Determines under what conditions, and to what extent and with what frequency, audits should be conducted in concert with the qualification process (by developing the relationship between costs of audit and expected losses due to	*a.* Enables the system manager to distribute administrative overhead dedicated to enrollment audits with respect to expected value of audit investments for: 1. Client classes 2. Access Modalities 3. Access Agents

Table 3.1 *cont.*

II. QUALIFICATION SUBSYSTEM

Instrument	Formulation	Implication
	fraudulent enrollments).	*b.* Audits are rationalized as to both subject and frequency, such that the program may eventually converge on an overall access-qualification structure and process that produces a given level of client population for the least associated expenditure of administrative overhead.

III. COVERAGE ASSIGNMENT SUBSYSTEM

Instrument	Formulation	Implication
1. Decomposition Function	Using primarily instruments of qualitative analysis (at least in the initial stages of program implementation), these functions define the different modalities by which services may be delivered to the client population.	*a.* In conjunction with the stratification functions (below), allows the delivery system to mount as many unique delivery or coverage modalities as there are unique clients and/or conditional classes. *b.* Different coverage modalities imply both structural and functional (processual) differentiation, and thus permits the system manager to place specific clients in the specific coverage modality which —given their attributes— promises to provide a local optimum with respect to cost-effectiveness. *c.* Allows the system manager to make a local determination about the point of best compromise between system efficiency as a function of a monolithic delivery process, and effectiveness as a function of differentiation.

Table 3.1 *cont.*

III. COVERAGE ASSIGNMENT SUBSYSTEM

Instrument	Formulation	Implication
2. Stratification Function/Congruence Vector	Subdivides any client population into unique socioeconomic (behavioral) and conditional (casualty) classes; in short, these stratification functions develop *typologies* of client attributes. These typologies are initially subjective in origin and subsequently (through a Bayesean "learning" process) become translated into empirically predicated constructs.	*a.* Permits the system manager to be both precise and flexible in the assignment of clients to the particular coverage or delivery modality most *congruent* with their properties. *b.* The association of different coverage or delivery modalities with different client and conditional classifications injects the opportunity for effectively optimal cost-effectiveness at the strategic or aggregate level. *c.* Permits the system manager to allocate overhead services to clients (e.g., guidance, tracking, referral) on the basis of their demonstrated need, rather than on the basis of a single set of system support being given to all clients; moreover, it distributes bureaucratic demands on the clients rationally rather than arbitrarily.
3. Density Distribution Regulation Functions (and Centrix Models)	Show the relationship between client clusters and provider locations (in terms of both geographic and intensity factors).	*a.* Allow the system manager to rationalize the placement of block providers (public service outlets), relative to projected demand and current provider configuration. *b.* Provides the basis for the system manager to "encourage" mainstream providers into certain areas where coverage gaps are most severe (e.g., provide for the rationalized location of

Table 3.1 *cont.*

III. COVERAGE ASSIGNMENT SUBSYSTEM

Instrument	Formulation	Implication
		HMO's and new physician private practices).

IV. REFERRAL SUBSYSTEM

Instrument	Formulation	Implication
1. Provider Match Algorithm	Schedules referrals for treatment to that provider which emerges as most favorable on the following dimensions: 1. Specialization 2. Geographic proximity to the client 3. Cost-effectiveness index 4. Number of previous referrals (an inverse factor)	*a.* Attempts, at the operational level, to ensure consistently adequate effectiveness (quality of service) while economizing on costs of treatment. *b.* Can be used to prevent the emergence of exploitative providers (e.g. Medicaid mills). *c.* Distributes program-related referrals in such a way that an effectively optimal client-provider configuration eventually emerges. *d.* Serves to provide impaired or inexperienced clients with specific directions about appropriate providers (and also allows control of the average cost per contact by program authorities).
2. Condition-Treatment Correlation	Permits the development of provider effectiveness indices by associating with each unique condition (diagnosis) treatments divided into the following classes: 1. Necessary (imperative) 2. Desired (optional) 3. Gratuitous (of doubtful utility) The correlations may be either historical (empirical) in origin or be predicated on normative or	*a.* Allows the program to rank providers in terms of their quality (effectiveness). *b.* Allows the system to serve client interests by flagging cases of undertreatment, and serve the cause of efficiency by penalizing providers for consistent overtreatment. *c.* The provider effectiveness index is adjusted downward for either of these faults, and this makes further

Table 3.1 *cont.*

IV. REFERRAL SUBSYSTEM

Instrument	Formulation	Implication
	judgmental criteria (via a Delphi process, for example).	referrals successively more improbable (via the provider match algorithm). *d.* Offers the system manager an opportunity to adjust (ad hoc) level of service the program will provide clients.
3. Standard Cost Distribution	For unique treatments or treatment sets, establishes three intervals into which costs may fall: 1. Significantly lower than average 2. Average 3. Significantly higher than average As a rule, the cost continuum will be developed from historical information, either local or analogic in origin.	*a.* Allows development of a cost index for each provider, based on the transactions for which he has requested reimbursement at a certain level. *b.* Allows eventual control of economy of aggregate service by making relatively more referrals to the more continent providers in any given specialty and/or geographic area. *c.* Allows projection of resource requirements for categorical funding sources. *d.* Allows a priori adjustment reimbursements to providers when requests are beyond tolerable limits.
4. Provider Cost-Effectiveness Indices	Develops, for each provider, a cost-effectiveness index that will be used to determine the probability of a referral being made via the provider match algorithm.	Allows the system manager to allocate resources in disaggregate so as to obtain an effectively optimal investment of direct (nonoverhead) funds.

V. INTERNAL (FISCAL) CONTROL SUBSYSTEM

Instrument	Formulation	Implication
1. Expenditure Extrapolation	Projects the rate at which a particular fund (categorical or general account) is being	Allows the system manager to protect the integrity of the funding base by adjusting system

Table 3.1 *cont.*

V. INTERNAL (FISCAL) CONTROL SUBSYSTEM

Instrument	Formulation	Implication
	exhausted.	parameters in light of rates of expenditure, relative to temporal or capitation budget limits.

2. Referent Demand Function	Allegorizes the schedule at which clients are expected to appear for service, and the expected number of service contacts, etc. The functions may be either of the following:
	  1. Induced-Empirical: based on historical projections
	  2. Deduced: based on theoretical models (e.g., learning-curve; Poisson)
	These curves are used to derive resource requirements and to suggest normative expenditure rates for each fund (or for unallocated accounts).
		a. Allows the system manager (and central funding authorities) to rationalize resource requests and aggregate allocations.
		b. Permits the re-allocation of resources within funding or fiscal periods.
		c. Permits the system to develop intermediate range plans based on projected changes in the client population, casualty conditions or changes in policy parameters.

3. Information Leverage Function	With reference to computations of *expected value of decision error,* attempts to impute a value to increments of information to be collected and manipulated.	*a.* May be used to rationalize reporting requirements by relating each to a specific decision function. *b.* May be used to generate cost-benefit indices for components of the system data base and to determine the rationalized structure of internal reports. *c.* In conjunction with internal and external reporting functions, may be used to develop the relationships between the management and planning functions of the system.
4. Inflation Adjustment Functions/Localization Indices	Shows the way in which the local economy adjusts to the projected inflow of social service funds or supports.	*a.* May be used to modify centralized estimates of sustenance requirements to local conditions (e.g., acts to inflate or deflate

Table 3.1 *cont.*

V. INTERNAL (FISCAL) CONTROL SUBSYSTEM

Instrument	Formulation	Implication
		the expected effect of centrally determined support levels when applied to a particular region or locality). *b.* Measures the extent to which the price of locally obtained services and/or goods responds to the increase in effective demand among social service clients, and considers the possibility that supply remains inelastic in the face of such incremental increases in demand (e.g., takes cognizance of the phenomenon of the ghetto economy). *c.* Becomes the central instrument in the development of a regional or "closed" macroeconomic model which can be used to project resource demands in light of local inflationary factors (and to show the effect of the social service "economy" on the mainstream economic system).

there will be an opportunity to amplify some of the arguments about the demands for managerial rationality.

STRUCTURAL ASPECTS OF THE SOCIAL SERVICE SECTOR

Initially, every social service program will have a certain client population with which it is concerned. Depending on the nature of the program, this client population may generally be thought of as a subset drawn from—or somehow differentiated from—some universe of individuals. This universe is the major referent for the social service program. It might, for example, represent the total population of some geographic region (a city, a county, a state, etc.), or it might represent some class of individuals (all women, all American Indians, all people over the age of 65, all individuals earning less that $4500 per year). Of course, geographic and class universes may be superimposed on each other such that, for example, a program may be responsible for all persons over 65 years of age living within certain city or county boundaries. Then class criteria may be superimposed as well so that a program may service all people over 65 years of age within the city limits who earn less than $4500 per year. Generally, not so much for purposes of real analysis as for purposes of program promotion, it is useful to define program client populations in simple set terms. In this respect, consider the Venn diagram in Figure 3.1.

Given these definitions of the basic subsets, the program population can be defined as the intersect of those individuals over 65 *and* those individuals earning less than $4500 per year $(X_\alpha = A \cap B)$. Similarly, the set D is called the program complement, being the set of all individuals who are under 65 *or* who earn more than $4500 per year: $D = U - (A \cup B)$.

In the general vernacular of the social service sector, any community or nation or region, etc., will have what is called a population of *social casualties.* The population of social casualties would be the set of all individuals who, for one reason or another, are included in the client population of some social service program. That is, within any geographic location, we would be able to define a set of subsets: $X = (X_1, X_2, \ldots X_m)$. Included in each

FIGURE 3.1 / Client populations viewed as subsets

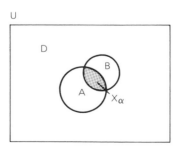

U = the universal set of all residents within a geographic boundary

A = the set of all individuals earning less than $4500 per year

B = the set of all people over 65 years of age

of the X_i subsets would be individuals with specific problems (e.g., the subset X_1 might be defined as the subset of all individuals within County Y of American Indian extraction; the subset X_2 might include all individuals who are heads of households but temporarily unemployed).

Now, as the reader might be aware, the history of social services in the United States (and in other nations as well) has been one characterized by the successive elaboration of client populations, by the continuous refinement of social casualty categories. The net result is that the number of unique subsets contained in X (the universe of all social casualties) has concatenated over the years to the point where there are now many X_i's, and therefore many social service programs dealing with social casualties of one type or another. When the concept of social services first became operational as a "managed" phenomenon, in Elizabethan England, there were very few (and for a time only one) sets into which social casualties might be placed. But today, looking at the formulation in the previous paragraph, the limiting subset X_m may have a value for m which exceeds 1,000. This means, in effect, that there are approximately 1,000 times as many social service programs today as there were when the concept of institutionalized social services was first initiated.

The resultant situation may only be described, and quite kindly, as an administrative mare's nest—a bureaucratic chaos. In relatively specific terms, this chaos presents itself in the following ways:

1. There will be cases where a single social casualty becomes the responsibility of more than one (and sometimes many) social service programs. That is, an individual may, for example, be blind, over 65, medically indigent and a subsistence farmer. In such a situation, he would have to apply, and deal with, four or five different agencies to garner the support to which he is entitled by various legislations. In terms of the simple set constructs, we thus get something like Figure 3.2. In practice, the incredible elaboration (or partitioning) of social service programs produces a competition among different agencies for clients having multiple casualty conditions, or at least places enormous bureaucratic demands (reporting, enrollment, etc.) on casualties with a complex of conditions.

U = the domains of all social service programs

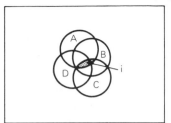

A = program for the blind
B = program for the aged
C = Medicaid/Medicare
D = program for aid to
    subsistence farmers

i = the individual in
    question

FIGURE 3.2 / Multiple conditions "map"

2. There will also be cases where the domains of two or more
social service programs may overlap, such that two or more
administratively distinct programs have essentially the same
responsibilities. When these agencies belong to different
bureaucratic groupings (e.g., one to the Department of Agriculture, another to HEW), there may be no coordination at
all between the agencies, and a resultant erosion of scarce
social service resources. A diagram of this situation would
look something like Figure 3.3. Here are two conditions of
redundancy, both quite widespread. In the case of concen-

FIGURE 3.3 / Redundancy of responsibility

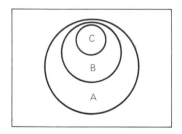

a. Concentric Systems: A = Federal Program
                              B = State Program
                              C = Local Program

Here, structure and function are essentially
replicated at three different levels, with
the "value added" at the two higher levels
possibly quite negligible

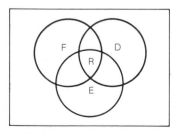

b. Overlapping Systems: D = Department of Agriculture
                              E = HEW
                              F = Department of Interior

Here, several different agencies all have
simultaneous authority over—or interest
in—a single area, defined at the nexus of
their several domains

tric responsibilities, the degree of redundancy is determined by the "value added" at the various levels. The *properly* concentric system would imply that there are significant differences of function defined at the several levels, such that the operations and responsibilities of the lower-level subsystems cannot be induced from those of the higher-level subsystems. In the social service sector, however, this is not the case. There is strong similarity of both structure and function among agencies operating at the several levels, and this implies a significant degree of redundancy. This, in turn, would suggest that the operations of one or more of the levels are essentially gratuitous. In the case of the overlapping systems, we get the situation where the domains of responsibility for several different systems are defined so as to yield a "populated" interface (or intersection). This means that the social service sector is not itself structured in terms of mutually exclusive responsibilities, and in practice gives rise to the possibility of functional confusion, interference and anomaly (where two different programs are pursuing different ends within the same sphere of influence). This problem becomes very much more serious when the overlapping subsystems are responsible to essentially different suprasystems, such that there is no nodal point of central authority. In the example, Interior, Agriculture and HEW are all departments of the executive, but the executive does not have the managerial technology to regulate their interface or rationalize their relationships. Therefore, redundancy goes virtually unchecked, with the resultant erosion of the cost-effectiveness of the executive in aggregate, and occasional chaos at the operating levels where collisions of interest take place.

3. There is, of course, another sad corollary to the bureaucratic elaboration of the social service sector. This is the case where certain problems tend to fall neatly between the interstices of the various program domains, such that certain casualty conditions remain undefined and therefore untreated. This happens more frequently than supposed, as leads to a set of "catch 22s" that operate throughout the social services sector. Just by way of example, suppose there is a white male, over 21, who loses his job because of technical unemployment (e.g., the skill he had developed has be-

come obsolete). He might very well receive no training as in Figure 3.4. So, because the individual is neither an Indian, a minority group member, handicapped or a juvenile delinquent, he is out of luck.

The point to these various arguments is simply this: the very structure of the social service sector, viewed in aggregate, imposes certain built-in inefficiencies and instances of ineffectiveness (these latter being defined by the lacunae among the various social service responsibilities, as in the last example). One might very well ask how could this unwieldy, virtually indefinable structural melange have come about? The answer is relatively simple. We have to recognize that the development of the social service sector was based on an ad hoc process. Therefore, there was no opportunity to develop an overall, rational structure to begin with. This ad hoc process is fueled by three separate subprocesses. First, there is the fact that congressmen see their function as the elaboration (or the parochialization) of the legislative base, not its simplification or rationalization. It is a natural tendency of congressmen (perhaps reflecting their predominantly legalistic orientations) to be constantly alert for opportunities to introduce new bills and gain the affection of new subpopulations. That is, certain congressmen tend to become champions of certain causes and gain their identity thereby. As one cause is exhausted, a new one must be created or responded to. Thus, the social service sector reflects the constant concatenation of the general legislative base, and this constant concatenation (and therefore complexification) merely reflects the interests of the congress itself. Second, there has been a complex change in the

FIGURE 3.4 / Lacunae among social service coverages

U = the domains of all social service programs

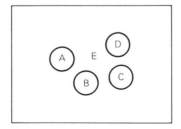

i = the individual in question

A = program to retrain American Indians

B = program to retrain the handicapped

C = program to retrain minority youth

D = program to retrain juvenile delinquents

E = excluded set (the proportion of the population not covered by any retraining program) attitude toward social services, a change that owes some portion of its substance to the recently redefined interchanges between social, economic and political processes. Basically, social services were once characterized as institutionalized charity and were held to be a privilege rather than a right. Now, the concept of rights has pervaded virtually every sector of our nation and with it come agencies to protect or further these rights. Certain natural processes have also served to expand the social service aegis, particularly the fact that employment opportunities are growing scarcer with respect to the employable population, the fact that the welfare process becomes a syndrome of entrapment (if not enticement) and leads to its own autochthonous expansion, and the constant attempt to raise the poverty threshold, etc. Third, there is the tendency of bureaucratic agencies to constantly seek to expand and elaborate their purview. It is something of an axiom that a businessman makes more money the fewer employees he has, whereas a public administrator's salary is geared to the size of his staff. This means that social service managers are always eager to define and add new responsibilities and expand their domain of influence. In this regard, some agency heads support congress in the elaboration of new legislation (often suggesting gaps that might be filled or new programs that might be mongered). In other cases, agencies might fight tenaciously to get control over new ventures. But the prevailing and probably most potent force is the tendency of the social service bureaucrat to consider that elaboration of the administrative structure is the same thing as the amplification of agency effectiveness.

All of these structural defects are present, to some extent, in the existing social service sector. The net effect is not simply the confusion of responsibilities and organization which amuse daily columnists and confound conservative critics of social service programs, but also the direct and significant dilution of the "effect" of funds transferred to the social service sector. Again, then, managerial errors and ill-informed policy decisions return to haunt the social casualty and blunt the impact of the programs in question. The critical issue here, however, is that there does not really appear to be a serious effort afoot to remedy the structural defects. Rather, the processes which led to the elaboration ad absurdum of the social service sector—and which have forced a concatenate increase in aggregate administrative overhead—continue largely unabated and unchallenged. Among these processes are the following:

1. The blanket centralization of social services at the federal level, without any accompanying attempts at meaningful integration

2. The tendency to associate a new administrative apparatus with each new category of social casualty defined

3. The acceleration of the definition of new social casualty classes as an adjunct to "egalitarian" sociopolitics

To most operating social service functionaries at the local level, these are familiar problems. They all stem, basically, from managerial failure at the highest level . . . from the fact that the social service structure has evolved ad hoc, and that the resultant complex of programs exceeds administrative and rationalizing abilities.

This ad hoc social service structure stems from several sources. Initially, system analysts and management theorists must bear part of the blame; they have tended to suggest that centralization and integration are indeed possible, without effective limitations. Thus, some government functionaries have come to believe that the social service sector should be organized as a complex of interests and agencies where central authorities control not only broad policy issues, but also maintain day-to-day (or at least period-to-period) control and regulation over all facets of operations. In this assertion, they have taken the lead from the many system analysts whose enthusiasm tends often to outstrip their genuine capabilities. The bald fact is that the managerial state-of-the-art does not as yet permit rational management of allocations or relations for extremely complex, widespread and protean systems. Thus, as new units and elaborations are added to the basic system structure, they cannot be well or fully integrated into the complex, and the net result is an organizational complex that looks something like the old Jaguar automobile engines, where the fuel pump worked through the radio circuitry and where it was virtually impossible to comprehend how the motor worked simply by referring to blueprints or design criteria.

It is to be expected that most field managers recognize these managerial limits, even if some may be enamored of the promises made by flashy system theorists and grand consultants. At least some bureaucrats at the federal level are alert to the fact that the current social service structure is an olio that only Rube Goldberg could appreciate. But their apprehensions are overshadowed by elaborative pressure from another direction altogether. Even as the administrators cry halt to the elaboration of the social service structure, members of Congress are constantly adding new apparatus and funding new categorical programs. To some critics of congressional processes, it appears that the opportunity to define a new social casualty class—and hence derive a new constituency—is a temptation congressmen cannot resist. But once defined, this new class becomes a de facto elaboration of the existing structure, and it often comes equipped with its own legislatively determined (and protected) administrative structure. Thus, managerial prerogatives are often constrained by explicit legislative actions, such that even well-intentioned efforts at rational reorganization are a priori estopped.

But one must appreciate the pressures under which the Congress operates. Particularly, ours is an age where the population at large has helped transmogrify the government from a coordinative body into a sort of gigantic gum-ball machine; anyone with the proper political coin, who pulls the right parochial lever, finds himself the recipient of a happy surprise in the form of some dedicated, categorical subsidy. The very rich and the very poor alike operate the machine, and hence the government's guaranteed constituency—its beneficiaries—constantly expands. And because the social service sector is the mechanism for delivering political largesse, it is forced always to broaden. Then the bureaucratic preference for a new administrative unit for each category—which is both the simplest organizational strategy and, as suggested, occasionally the strategy dictated by Congress—kicks in and the organization chart must be redeveloped to show the new members of the family. And because this elaborative process is so constant, time itself combines with technical limitations to defy actionable integration and rationalization.

It is an axiom of serious system theory that the ability to integrate or rationalize a system varies with its size (its domain), the nature of its constituents and the dynamics of the milieu in which the system is resident. That is, a small system, resident in an essentially placid environment and comprised of essentially replicative units is the easiest to rationally manage and integrate. The independent variable is, of course, the sophistication of the managerial technology. But as already suggested, there are severe limitations here. The implication, therefore, is that both the tendency toward centralization of the social service sector and the tendency toward elaboration via categorical programs frustrate rationality and preempt integration. Most management theorists—and all serious system scientists—would thus suggest that the therapeutic strategy would be to investigate the comparative advantage of a decentralized, catholic social service structure. This means, in effect, that management logic would argue for the concentration of social service prerogatives at the local level. As local systems would be significantly constrained in terms of domain of influence, environmental complexity and number of functional constituents, we should be able to use existing technology to force integration and rationality, whereas extremely massive centralized systems consistently defy the repertoire of analytical instruments. This does not imply that local systems are easy to integrate or rationalize; rather, it simply suggests that they should be *easier* to properly administer than state or federal systems. And we, as either managers or system designers, are not in a position to deny ourselves any increment of simplicity.

But there are some popular arguments against localization or decentralization which are not entirely persuasive. First, there is the hackneyed assumption that local government is more susceptible to corruption and subornation than state or federal government. Recent events have suggested that this assumption may be misplaced. Second, there is the legitimate suggestion that the benefits of centralization would be foregone through delegation of social service authority, and that these lost benefits would outweigh the gains in efficiency that decentralization would promise. Even the basic theoretical model that would enable justification of such an assertion has not been developed. Moreover, there is the practical matter mentioned earlier: the benefits of centralization in the social service sector imply a level of managerial sophistication that simply does not exist (and even that technology which is immediately accessible to operating functionaries is seldom exercised). In short, centralization, in the absence of administrative adequacy, merely yields all the inefficiencies without generating the facts of greater effectiveness and rationalization. Finally, there is a third factor: the deep-seated and pervasive derogation of state and local rights and the acceleration of federal prerogatives. This is fueled by the system by which taxes are collected and preempted, and softened only partially by movements toward revenue sharing and away from categorical supports. Associated with this movement toward concentration of prerogatives is a far more subtle sociobehavioral process: the assertion that the ends of the nation would be well served by attempts to eradicate local differences in favor of a distributed symmetry of socioeconomic properties. Many academics and social critics have apparently become convinced that individual communities should look pretty much like one another, and that differentiation—on virtually any dimension—is a prima facie case for discrimination. In this assertion, they have had limited support from federal district courts (though not much from the Supreme Court), and the vocal and obstreperous affection of what might be called the "ultraliberal" factions. Thus, racial equality, sexual equality and affirmative action programs have sometimes been used as a source of justification for the removal of all signatures of socioeconomic asymmetry . . . for an end to all zoning restrictions, for the centralization of support for schools and the dethroning of local school boards, for compensatory quota systems in education and employment, for progressive tax rates that approach confiscatory levels. Taken all in all, it sometimes seems to critics of such efforts that they are designed to affect sameness by reducing the level of welfare allowed the advantaged segments of the population, not by raising the effectiveness and performance parameters of the disadvantaged. To the extent that a program is essentialiy just a fiscal transfer mechanism—or rooted in a maintenance rather than developmental posture—this indictment probably has some merit. On a broader level, it may be well to reflect that symmetry of prerogative and sameness of attributes are characteristics of the most primitive societal systems, not the most advanced. Therefore, the support of centralization as a vehicle for inducing sameness may not be a particularly sophisticated position to hold, even if one could presume that symmetry can indeed be legislated or forced on a people.

There is a final argument for centralization. This is the suggestion that a strong central bureaucracy is required to provide the ongoing research and consultative activities required to support local programs. The theory here is good, but a glance at the schedule of research and consultative activities supported inhouse by the various central federal agencies yields some disappointment to those looking for innovation and developmental ingenuity. Indeed, the federal agencies seem to confuse advice and consultation—and research—with the generation of new reporting requirements and controls, as many local social service managers are quick to point out. Moreover, there appears to be no monopoly on insight or talent among the professionals employed by the bureaucracies; rather, the majority of proper research and development—both substantive and managerial—is still done by state and local universities and by the occasional private research house. Therefore, once again, the federal bureaucracies seem to be more of a simple conduit for funds than a sterling and irreplaceable source of technical support.

In summary of the structural matters, then, there do not seem to be any irrefutable defenses for centralization of social service functions in the absence of a truly responsive management technology, and there are even fewer defenses that could be generated for the perpetuation of the currently popular basis of categorical (parochial) funding procedures. Neither centralization nor categorization—as currently practiced—secure the integrity of social service programs nor protect against the abuses that some individuals ascribe to local managements. In short, we bear the added expense of concentricity and redundancy of structure without really benefitting from the advantages that could result from effective integration at successively higher levels of the social service system. More seriously still, it seems that those very functions which should be centralized at the state and federal level—and which should be explicitly integrated on a broad basis—are those that still remain essentially under local control. Particularly, there are basic constitutional arguments for the centralization of education, for example, and a practical argument for the regional or federal control of air and water pollution standards, etc. The local school board does represent a threat to the constitutional implication for equal opportunity (as the recent decision to use statewide rather than local funding for education in New Jersey suggested). Pollution, in its right, does not obey community or state boundaries, and to leave effective control of this function to localities or narrowly defined agencies violates logic if not the law. To a certain extent, the same thing is true of general welfare programs. Here the states and localities largely support the programs, but the indigent or casualty population is not immobile. Thus, indigents will migrate to a locality where a particularly attractive social service system has been established, as happened both in New York City and more recently in Denver. The influx of indigents thus erodes the local population-resource base and the originally attractive system quickly becomes inadequate. Yet the federal system does not compensate the local system for the migrants, and the consciona-ble, responsive community is thus penalized for its success. Add-ing insult to injury, it seems, is the federal government's refusal to allow local programs to protect against such erosion through the development of residency requirements or other specific qualification criteria (which would supplant the generalized en-rollment qualifications established by the centralized, categorical funding sources).

Without going into any further detail on the matter of struc-ture, it should be clear that the rationalization of the social ser-vice sector demands not only new technological efforts in terms of direct management, but also some major modifications on the structural dimension. Those that seem particularly appropriate for consideration—qua hypotheses—are the following:

1. Curtailing the size and authority of the federal bureaucra-cies that in effect perform only a conduit function at grossly inflated administrative (transaction) costs. They should be reduced to a cadre sufficient to support the basic transfer functions, and to actively control only those few programs that logic or the Constitution dictate should be centralized.

2. Curtail those state and regional agencies that are part of the concentric pattern, and serve merely to impose barriers or convolutions along the conduit from the resource system to the local social service program. This strategy, in association with that above, would serve to reduce redundancy and increase responsiveness.

3. Gradually eliminate the categorical basis of funding and ad-ministrating social service programs, replacing categorical grants with block grants or expanded revenue-sharing flows. In the process of diluting the categorical base, it should be possible to gradually eliminate those areas of redundancy due to functional overlap among different agencies and de-partments.

4. Establish rotating committees drawn from academics, the lay public and professional personnel to determine funding priorities for research and development grants, and to de-termine what demonstration projects should be supported in what communities. If possible, these committee members should serve on a voluntary rather than remunerative basis,

thus eliminating the enormous overhead associated with the process-oriented transfer operations now seated in the central bureaucracies.

5. The demise of the categorical basis for elaborating and funding social service programs—along with the acceleration of the revenue-sharing process—would permit local authorities to determine coverage decisions. Participatory working committees could determine on the basis of need and demonstrable effectiveness what casualty conditions to support; the system manager would then, on the basis of his fiscal considerations, determine the intensity of support. Many other options for local determination of the social service structure are also available, most of which could be expected to represent an improvement over the process which finds program definition in the hands of congress and program parameters determined by some central bureaucracy.

6. Seat the basic reporting and accountability authorities in the hands of local select committees, with an ombudsman provision (e.g., control by exception) available on call at the regional, state or federal levels, perhaps through an adjunct to the system of standing courts. To the extent that *both income and expense* are tied to local communities—accessible to their citizens and constituent interests—managers may be expected to be perhaps more responsive to matters of integrity than is the case with the currently centralized (and largely unauditable) reporting and accountability system.

7. Finally, with the easing of centralistic and categorical constraints on local programs, there is the opportunity to inject effective integration, as managerial limitations would become less strict with the narrowing of the domain. Combined with the above strategies—the curtailment of intervening bureaucratic structures, the voluntary control and determination of functions and the limiting of the partitions dictated by categorical funding—integration at the local level promises to make significant improvements in the proportion of aggregate dollars going to the social service sector which find their way into directly productive investments.

As was suggested, however, these are just hypothetical strategies. A formal consideration of their implications might lead to the situation where some are more attractive than others. But all, to some extent or another, are the type of basic strategic or policy issues that must be attended to before improvements in local management will result in improvements in the social service sector in aggregate.

It is necessary to undertake one important qualification in the presentation of these structural strategies. This is a definition of what is meant by *local* management. It must, of course, be defined very broadly. Initially, local management might be coextensive with a particular community or a political domain, township, county, city, etc., but there are many units of local government that are simply too small to be the seat for any kind of formal social service program. To tie these small localities to other political units—to take all the small independent communities surrounding a large metropolis and force them onto that city's structure—might offend both the smaller towns and be an unwelcome burden to the city's managers. Therefore, a local management in this case might be a regional authority of some kind. Clearly, for rural areas, a regional authority would also be dictated. In still other cases, local management might simply imply a delegation of authority from higher-level bureaucrats to lower-level, community or region-based bureaucrats; such managers would be local in terms of their orientation, but federal or state in terms of their basic organizational affiliation. Thus, local social service management is not the same thing as local government; in short, the localities defined for purposes of social service administration need not be the same as the localities defined by political boundaries (and, indeed, the H.S.A.s that have recently been defined by HEW do not follow such boundaries).

This qualification immediately raises another: what are the comparative advantages of centralization and decentralization in general? This is a question that has consumed much space in management literature, and we may dispose of it here rather quickly by employing a logical artifice. Particularly, the advantages of centralization are found, as a rule, on the efficiency dimension: (a) the elimination of redundancies; (b) the economies of scale associated with certain logistical functions and procurements; (c) the extension of the zero-opportunity cost criterion across a wider domain of investments (which means, in effect, that a central decision scheme applies to a significantly large number of programs, such that a larger quantum of resources is rationalized in one simultaneous process). In more general terms, the advantages of centralization are intimately tied to the extent to which concentration of managerial and fiscal authority yields a degree of *integration* and *synergy* among a number of different functional units. In contrast, the advantages of decentralization appear theoretically minimal; about all that can be said, with any assurity, is that delegation of authority can improve the probability of a system being effective, providing that the function it performs (its mission) is sufficiently complex, and providing that the environments over which it elects to operate are sufficiently protean. When these two conditions hold, both *expediency* and *sensitivity* are served, but only at the expense of the quasi-mechanical efficiency that can be programmed into strictly centralized systems. Because the social service sector is generally characterized by complexity of mission (at least when we rise beyond mere maintenance) and protean milieux, decentralization gains some theoretical advantage as the structural modality for social service systems.

There is an even more practical and direct argument for decentralization in the social service sector, one repeated frequently: the current state of management technology (as practiced or as immediately available) is not sufficiently sophisticated to permit large-scale, centralized systems to achieve the two advantages of centralization to any significant extent. That is, for extremely large-scale systems, we simply do not have the analytical skills to inject truly effective integration or realize truly substantial synergy. Thus, beyond the theoretical arguments raised above, there is a practical constraint on the degree of centralization which can be supported given in the level of management technology accessible to the system managers (and the level of technology at which they are able to operate due to their own limitations). The many horror stories that emerge about lack of sensitivity in the social service sector—about redundancies, fiscal abuses, operational errors and dramatic dysfunctions—are ample testimony to the fact that centralistic bureaucracies have exceeded optimal scale of plant (not necessarily from the standpoint of economic criteria, but certainly from the standpoint of managerial technology). And in the long run, economic and managerial constraints become virtually indistinguishable in terms of the derogatory effects on the aggregate cost-effectiveness indices an organization can obtain. Thus, the urging that we consider some active decentralization is predicated on both theoretical and pragmatic grounds, and the critical qualification is that as management sophistication increases, so will the appropriate degree of centralization.

The basic thrust of the structural strategies just introduced thus urges movement toward decentralization by the mechanism of increasing *delegation* of authority from centralized to local systems. Revenue sharing, as opposed to categorical funding, implies a significant degree of delegation, as does the suggestion that accountability and reporting functions be localized (subject to the broad definition of local developed earlier, and excluding those few functions—environmental protection, education, etc.—that are better centralized because of constitutional or logical considerations). The rationale for the strategies is dualistic, as developed above: it recognizes the predominance of responsiveness and expediency (the dimensions of effectiveness) over efficiency criteria, and also takes cognizance of the more telling factor, technological constraints.

There is a third factor involved with the matter of restructuring the social service sector—ideology. There are many who suggest that central bureaucracies *cannot* delegate authority over social service functions because the authority does not rest with the central government, but with individual localities or communities. The rationale supporting this argument (heard a bit more frequently these days than in the decades immediately past) is this: call it by any other name, but maintenance-oriented social service programs are still just institutionalized charity. And neither the Constitution nor democratic sociopolitical precedent yield authority over charity to any governmental unit, much less the federal government. Therefore, according to this argument, the government has no authority to preempt resources for transfer payments, and certainly no authority to impose schemes like social security or national health care on the population at large. The argument ends, then, with the demand that social services be thoroughly decentralized, and that the taxes that go to support social service functions be retained locally. Thus, even revenue sharing is considered an insult with the federal government presuming to do communities a favor by returning resources that should not have been collected in the first place. Now, this argument is not simply being raised by rock-ribbed reactionaries, but also by many who could only be considered political moderates.

The fundamental implication they offer is that while protection of equal opportunities is indeed a right guaranteed by the Constitution, access to charity via transfer payments is at best a privilege vulnerable to the whims of national conscience and a variable axiology.

Now, not being able to completely forecast which way the ideological or axiological winds are going to blow, it behooves the social service system designer to inject some flexibility into the systems he designs. That is, the technical characteristics programmed into the systems being designed now should, wherever possible, be appropriate to either of two antonymous structural states which might possibly emerge: (a) more intense centralization of social services in company with increasing sophistication of system management instruments; or (b) decentralization and decategorization of programs in response to a retrenchment of conservative ideology, and the consequent reduction in the number of maintenance functions performed, and an increase in the number of distinctly developmental social service programs. In speculation here, it does indeed appear that most system designers and technical management specialists—at least those operating outside the political or ideological mainstreams—are indeed developing designs which are not dependent upon any particular politico-ideological context. This will become clearer as we now take a look at the attributes of a prototypical social service delivery system.

A GENERALIZED SOCIAL SERVICE LOGIC

The best evidence that a generic social service management science (or logic, if you will) might indeed be useful is the fact that those social service systems which have evolved all share certain characteristics in common. This is particularly true of those systems that could be described as products of proper system analysis efforts, design exercises where imagination did not too far outstrip common sense and technical state-of-the-art. If we were to abstract such systems, we might find a structure something like that presented in Chart 3.1.

Access Modalities

As the chart suggests, the front end of a social service delivery system entails the *access* function. This has the task of recruiting casualities into the program. There are several different access modalities, each of which may be used alone or in combination with others. The five described below seem to be the most frequently employed:

1. *Promotion,* whereby the provisions of the program are advertised and enrollments encouraged through the media ... a passive modality

2. *Search,* where staff personnel are assigned the task of seeking possible clients, often through a secondary research process (e.g., examining the files of other agencies; reviewing tax schedules for income levels; checking the police blotter or the registers of transient hotels), or often through some sort of direct canvassing "in the streets"

CHART 3.1 / General logic

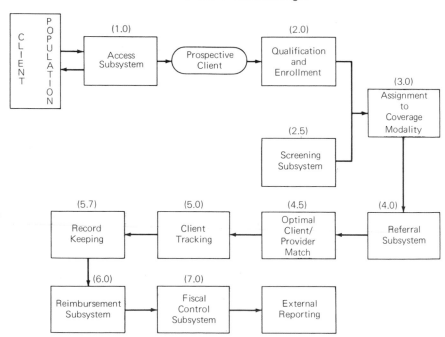

(Note: Index numbers refer to segments of the master logic charts included in Chapter 2.)

3. *Recruitment,* where various community officials (e.g., church pastors, boy scout troop leaders, police, doctors, public health nurses, rescue personnel, even utility representatives and other general business functionaries) are encouraged to actually forward candidates for enrollment

4. *Enlistment,* where there are offices accessible to those individuals who, on their own volition, wish to seek coverage by the program and therefore "access" themselves

5. *Transfer,* where other functionaries in the social service sector (case workers, juvenile and truant and probation officers, medical providers, etc.) provide the client with a cross-referral to the program

In all cases, the ambition of the access subsystem—or its performance criterion—is the proportion of all potential clients (the casualty universe, C) actually made aware of the program's existence and provisions: c/C, where c is the actual level of contact. Generally speaking, the managerial task is to generate what is called the greatest *access leverage,* which means getting the most contact for the program with the lowest associated expenditure of resources. Therefore, when a program is initiated, we have to consider the appropriateness of the various access modalities for the client population with which we are concerned and also their relative cost.

For example, take the matter of promotion. It is first of all necessary to make a strategic decision about the expected effects of using promotion as an access modality. Then it is necessary to make some computations about its expected leverage (efficiency). The strategic aspect may be handled with very casual logic. Particularly, promotion will usually yield better results when (a) the client population being persuaded is a stable, stationary one; (b) the population is literate and expectedly alert; (c) the casualty condition to be supported is nonpejorative (e.g., a relatively innocent situation); (d) the condition is reasonably widespread; (e) the program is unique in nature and concept; (f) the treatment the program entails is not likely to cause apprehension or discomfort. Thus one would not, for example, attempt to use promotion as the only means of attracting transient alcoholics to a local drying-out plant, or as the sole means of accessing a generally secretive and often elusive quasi-criminal population (say drug addicts). As for the economic or efficiency dimension, one would simply seek to establish what are called the *promotional elasticities* for the various media to be employed (television ads, magazine blurbs, etc.). The idea here is to distribute promotional investments such that we get the most effective exposure for any level of expenditure, an often complicated but usually feasible management science exercise, and one that would parallel the strategic (qualitative) task of choosing among the major modalities themselves (e.g., suggesting how much emphasis to place on promotion, per se, as opposed to enlistment or recruitment).

Instead of suggesting the general logical and economic properties for each of the modalities, the reader might find the summary in Table 3.2 more useful.

Again, the selection of a particular access modality—or of some set—is one of the first decisions the system manager must make. Yet it is clearly the responsibility of the system designers to make a preliminary analysis, such that the resultant delivery system entails the capability to invoke any of the relevant modalities on demand.

Qualification and Screening

The first section of Chapter 4 presents a detailed analysis of the access and screening and qualification logic peculiar to the HCDS system, and there makes the first attempt at providing the reader with detailed "templates" for delivery system construction. Here, then, is a very brief and general discussion about the qualification and screening function, staying within the confines of Chart 3.1.

Once an individual has gained access to a social service system, the natural next function is to evaluate whether or not he meets the requirements for admission (enrollment). In most cases, a social service program will impose both conditional and socioeconomic constraints. This reflects the categorical basis of most existing social service programs, but also will serve to isolate appropriate clients for an integrated system. The only difference between the categorical and integrated systems would be that, in an integrated system, client screening and qualification need be performed once, at the initial point of access. In a set of categorical programs—lacking the articulative or linkage mechanisms—each individual program must perform its own parochialized qualification and screening tasks.

TABLE 3.2 / Summary of the Properties of the Several Access Modalities

Modality	Effectiveness Considerations	Economic Attributes
Promotion	Where the condition to be treated is widespread and nonpejorative, and where the client population is expectedly alert to the media	Most expensive in terms of direct dollar outlays (above-the-line expenses)
Search	Where the client is likely to be impaired or inalert, or where the condition is likely to be treated as pejorative (and perhaps even illegal) or otherwise embarrassing (e.g., VD or drug addiction); or where the condition may cause a panic situation but is currently isolated (e.g., a well-confined outbreak of, say, botulism)	Most expensive in terms of man hours expended per contact (most erosive of below-the-line budgets)
Recruitment	Where the condition is likely to be one unrecognized by the casualty (e.g., retardation, psychosis, emergent alcoholism) or where the client population is unlikely to initate access themselves (perhaps because of pride) or respond to any promotional effort (as might be the case with certain immigrant groups); also important in those cases where preventative or preemptive efforts are valuable	Least expenditure of direct program resources (makes maximum use of community volunteerism and donated initiative)
Enlistment	Where the client is likely to be able to act in his own interests and initiate a contact (used most often in conjunction with some sort of promotional effort)	Expensive in terms of the overhead required to maintain the contact or access offices (e.g., rent, clerks, utilities)
Transfer	Where the condition the program seeks to treat is likely to appear in conjunction (be clustered) with conditions currently treated by other programs	Involves a pro rata assesment for the general (aggregate) distributed overhead, perhaps shared by a large number of different community agencies

The qualification and screening tasks, are, conceptually, very simple: collect data on the individual's relevant properties and match them against the *threshold criteria*. The threshold criteria simply delineate the conditions for enrollment: the income level which would qualify the individual; the ethnic or racial requirements for enrollment, if any; the nature of the casualty condition(s) the program demands, etc. But there are some important ancillary issues. Initially, there is the question of the necessity for *auditing* enrollment applications, the business of testing to see whether or not an individual's application is a valid representation of his socioeconomic and/or conditional situation. A well-designed program will, of course, consider the matter of the necessity for audit in light of the different access modalities. For example, when an individual is accessed through the search modality, the agency or program may presume that there is little opportunity for the individual to fool us as regards his true attributes. Therefore, the rate of audit would probably be lowest for those individuals brought to the system by agency or program search personnel. To a certain extent, the same thing is true of the recruitment modality. When the local priest, probation officer, police or other community functionaries—presumably in a position of trust—promote an individual for enrollment, the agency officials may generally presume that the information provided by the recruiter is correct. However, there isn't quite the same degree of assurance as with the search modality.

Therefore, some type of audit procedure would be advised. Particularly, there are two relatively high-order variables to look at. First, do certain classes of recruiters tend to have different accuracy records? That is, are priests more gullible than volunteer nurses? Are police better judges of an individual's true situation than, say, members of the Junior League? Therefore, some cases associated with each of the recruiter categories might have to be sampled. By sampling is meant following up the information given on the enrollment request (or on the access form to which the recruiter testified and may have himself prepared) with a field inquiry. Next, note the *variance* between what the recruiters attested to and what the field researchers actually discovered about the client's qualifications. When this variance is consistently small for a particular class of recruiters, frequency of audit associated with that class is reduced. Were, however, variances strong and consistent for a particular class of recruiter, then the audit frequency for prospective clients proposed by that type of recruiter would increase.

The second variable is to assess the accuracy or validity of enrollments proposed by certain individual recruiters *within* recruiter classes. That is, we cannot presume that recruiter integrity is consistent. Therefore, we would probably want to audit certain cases submitted by individual recruiters and again assess variance. The sampling frequency might be very small, and the assessment of individual recruiters might therefore be a relatively inexpensive process. It is important to get an immediate fix on any recruiters who are consistently casual or inaccurate in their assessment or testimony about proposed clients' real attributes. At any rate, after a reasonable number of such audits, it is possible to gradually restrict the range of authorized recruiters to those classes—and those individuals—who prove most consistently accurate and eventually reduce the costs of audit for the recruiter-accessed individuals to an effective minimum.

Essentially the same audit logic would prevail with respect to candidates for enrollment proposed through the transfer modality (those clients sent to the program from other social service agencies). Limited audits would be conducted to identify the accuracy characteristics of certain types of transfer routes (e.g., are public nurses more or less accurate in their screenings than, say, welfare case workers?). And within the various transfer categories, the performance of certain individuals might be audited. We would thus get a schedule of expected screening accuracy for the transfer modality similar to that generated for the recruitment modality, seeking always to minimize the frequency with which audits must be performed.

It is to be expected that the most complete audit would have to be done on individuals proposing themselves for enrollment, that is, on prospective clients coming to the program via the enlistment modality. Here there is no first line of defense as it were; access is unmediated by any prior authority or any ancillary screening function. Again, however, auditing would be on a sample basis, but using different dimensions. Among the dimensions to test would be the validity of enrollment information coming from (a) different age groups, (b) different ethnic or racial classes, (c) different geographic locales, (d) different socioeconomic profiles, or on any other basis for client stratification that might emerge. The ultimate aim of the screening audits is thus to establish those particular enrollment requests which are most likely to be fraudulent, and therefore allow eventual concentration of auditing resources in those particular areas where fraud or abuse is most probable. In short, the aim is to minimize administrative overhead (transaction costs).

Even from this very brief argument, some general prescriptions can be seen, and some glaring faults of most existing social service programs can be cited. First, of the various access-screening modalities, probably the most economical—given expected costs of access, fraud and audit—is recruitment. The transfer modality is also attractive, to the extent that the costs of supporting agency interchange do not become excessive. The least economical way to invoke system enrollment is through enlistment; it implies expensive access overhead, and also the highest frequency of expensive, time-consuming audit in order to maintain any desired rate of accuracy. But the central fact of the access-screening attributes of most existing social service programs is this: they rely almost exclusively on the enlistment-promotional modalities and therefore have built-in inefficiencies right from the very initiation of the program. Because they have such limited resources they can devote to the audit function, the existing programs also have a very high rate of fraudulent enrollments. Therefore, in terms of a general prescription, social service programs should try to shift the burden of access and screening as completely as possible to the recruitment and transfer modalities and try to minimize the accesses sponsored via enlistment. In this way, the direct costs of access are minimized (that is, distributed out to the broader community in such a way that no recompense would be required), as are expected audit-fraud expenses. Thus, administrative overhead may be significantly reduced, and the resources potentially available to the client (dollar transfers, case worker support, etc.) are correspondingly increased.

There is another important aspect to the qualification-screening process that is seldom made explicit. Most existing social service programs—established under the parochial, categorical logic which currently prevails—do not make proper use of the *qualification threshold as a management tool.* * Rather, qualifications tend to be fixed by central legislation, and the individual social service programs thus once again become victims of bureaucratic constraints that appear to ensure equity, but may often just promote ineffectiveness and inefficiency. We shall not get the full implication of the variable qualification threshold until we have been through Chapter 5 but set out here is the rudimentary logic. The categorical (formula-based) funding mechanism under which most social service programs operate provides resources on a projected per capita basis (the general

*Note here that there is an obvious need to protect against inequities, perhaps through the type of ombudsman logic developed for the HCDS system; note also that "fixed" thresholds, as currently established, have not really prevented local inequities as was their purpose.

term for this is *capitation* funding). But, again, there is the problem that neither casualty conditions nor client socioeconomic properties are consistent from region to region (nor are relevant personal attributes distributed symmetrically even within certain local casualty subpopulations). An even more critical problem is this: the schedule by which social service casualties appear for treatment or program support is often a complex function. That is, requests for coverage or service tend to cluster through time, often in response to local perturbations which are simply unpredictable (e.g., plant closings, weather conditions). Moreover, inflation rates and basic sustenance variables (costs of medical care, rents and grocery prices, etc.) are often variable with respect to localities. The net result is that capitation budgets tend to be expended at something other than a constant rate throughout any fiscal year. Now, we have already mentioned that the degree of effectiveness of any social service program is related to the per capita resources it can command. But when the drawing down of these resources is inconsistent throughout a period, so is the level of effectiveness the program delivers.

Therefore, the level of service the system is capable of providing to its clients (given a fixed capitation budget) is also variable from the perspective of the local system manager. In this respect, there are two different behavioral referents that might be exhibited by local social service functionaries. First is what might be called the "crusader" orientation; his commitment is rather completely with the client population, and he considers resource constraints either to be artificial or inhumane, etc. The second type of manager is the "conservative"; he is fiscally oriented and may often emphasize his fiduciary relationship to the community at large. Now, this crude dichotomy is not any sort of serious sociological construct. It is offered merely in consideration of this assertion: the rate at which the resource base of a social service system is exhausted, given a capitation budget of any fixed amount, will in large measure depend on the behavioral orientation of the system manager himself. In particular, the "crusader" will often tend to draw down on the resource base very heavily during the initial phases of the fiscal year and therefore be left with an effectively diluted per capita resource base for the latter parts of the budgetary period. On the other hand, the "conservative" will often tend to be reluctant to expend resources in the initial phases of the budgeting period; he may then be left with excess resources during the terminal phases of

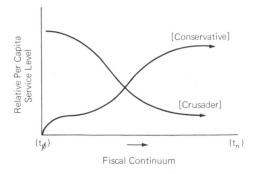

FIGURE 3.5 / Differential expenditure configuration

the fiscal year, which means in effect that the potential per capita resource level is amplified as time proceeds, for any fixed client population.*

Were a graphic reference provided for this matter of the rate of resource expenditures over a fiscal period, relative to behavioral orientation, we might get something like Figure 3.5.

The point, obviously, is that the quality of service (or the level of resources) delivered to system clients is not constant across the fiscal year, and the two different managerial orientations exhibit effectively inverse expenditure patterns. When either of these expenditure patterns is exercised to any extreme extent, dysfunctions develop from the client and fiscal standpoint.

Now, for social service programs which have a relatively long history of existence and relatively stable client populations, this problem may be dampened because there is considerable evidence on which to compute an expectedly constant level of resource expenditures. Whenever there is a potential for change in client population or in the economic properties of the community itself, perturbations become of potentially serious consequence and there may be significant discontinuities in the permissible rate of expenditures. This problem is amplified to the

*This situation leads to two seldom-mentioned problems. First, the excess resources are often not distributed in the form of enhanced benefits to system clients, but rather invested in conspicuous overhead—office expansions, managerial amenities, "junkets," etc. Second, when resources are to be expended on clients, the condition of scarce resources collapses and investments of excess resources are often made without any regard to cost-benefit criteria, thus lowering the overall performance index for the agency, and penalizing those clients serviced prior to the emergence of the funding excess. Again, the lack of a mechanism for rerouting excess funds makes either of these situations probable in the face of a "conservative" management.

extent that the program is a new one or an unprecedented one. If this problem is serious for the system manager working with a fixed capitation budget, it becomes even more serious for the social service manager whose budgeting base is variable or contingent. For example, some social service programs face essentially the same situation as philanthropic or private charitable operations: the budget may depend on the level of donations, on the yield from the local tax base or on the flow of exceptional grants (such as the resources made available through revenue-sharing provisions). As just explained, the confoundations that a variable budget may present to the manager of a noncapitated program are of the same type that population instability and economic perturbations present to the manager of a mature system on a fixed annual budget.

How, then, might the manager work to contain these confoundations? The obvious way is that just mentioned: *make enrollment qualifications variable instead of fixed.* If it is apparent that the rate of resource expenditures is insufficient (such that it appears a surplus would remain at the end of the fiscal year), the qualification threshold may be relaxed and more clients enrolled. The threshold could be tightened were resource rates excessive.* When some federal bureaucracy establishes fixed qualification criteria and then broadcasts these across all regions, it places tremendous burdens on the local social service managers and also deprives them of a critical control instrument. Threshold flexibility is perhaps the first line of defense against the perturbations that tend to cause such embarrassment to so many social service agencies. When qualifications are rigid, it appears that client populations are either complaining that they are being ill-served, or that the program is under attack from community critics as being profligate. Therefore, when the qualification threshold is not employed as a management tool—but fixed as a bureaucratic, centrally determined parameter—the already significant vulnerability of the local social service manager is increased.

---

*Or, as shall be pointed out in Chapter 5, the manager might cut back on promotion, retard other access modalities or perhaps institute some sort of triage mechanism. As Chapter 5 points out, the variable qualification threshold is not, of itself, a triage mechanism. For it may act to dampen breadth or intensity of coverage as well as to curtail the client population over some interval.

The Coverage Assignment Process

The logic of making assignments of different clients to different coverage modalities is an extension of the logic which caused stratification of client populations in the first place. In particular, we are simply extending the process of developing definitive *sets*. The next chapter treats the problem of coverage assignments in great detail (and actually shows the way in which assignments are made by the HCDS system). Here, however, the discussion may be very general.

The best example of the need for different coverage modalities—as opposed to a common method of treatment and control for all a program's clients—stems again from the joint conditions of effectiveness, efficiency and client dignity. In most existing social service programs, all clients tend to be treated alike, according to certain programmed bureaucratic procedures, irrespective of the individual client's attributes. That this sameness of treatment philosophy is still extant is best brought out by a fact mentioned much earlier. Particularly, when reviewing the various proposals for a national health care system, virtually all of them anticipate or consider just a single coverage modality: prepaid or insurance coverage. But as already suggested, this is a very naive strategy. Not all individuals are going to be capable of making rational decisions about what medical care they require and who is best qualified to provide that care. Again, the social casualty will differ considerably in his ability to make proper consumer decisions. Therefore, any properly designed social service program must take into account the fact that clients differ in their orientation, and it must be prepared to develop different coverage modalities that respond to those differences. If no such coverage distinctions are made, a system will emerge that is characterized constantly by ineffectiveness and incongruence . . . and a system which will be a vulnerable target for criticism from both the client populations and the community at large (members of the resource system).

Therefore, one of the first tasks of the system designers is to attempt to subdivide the client population into definitive subsets. In virtually all cases, there are two major dimensions which must be considered: (a) the sociobehavioral properties of the individuals; (b) the different conditional categories that present themselves among the client population. As the reader will see in the next chapter, both these dimensions were used to develop the client subsets and different coverage modalities for the HCDS system. Because the illustration of the partitioning process is fairly complete there (such that it could be replicated fairly easily by social service programs in other contexts), none of the technological arguments will be gone through here. Rather, here are a few words about the logical rationale that was employed in the HCDS design and which is generic to virtually all social service exercises.

Initially, it is important to maintain a parity between the number of different subsets into which clients are divided in terms of their sociobehavioral properties and the number of different coverage modalities that are eventually defined. That is, *each client subclass should be associated with one and only one coverage modality.* Thus, either the coverage modalities or the client characteristics may be the leading factor in the system design effort. In some cases there may be only so many coverage modalities which may be legitimately defined; these, therefore, may dictate the number of different sociobehavioral classes into which the client population is divided. In other cases, client characteristics may be significantly clustered, such that class distinctions beyond a certain number become blurred and trivial. In this case, the number of client classes would dictate the number of coverage modalities to be developed.

The point, however, is this: by subdividing the client population and defining correlative coverage modalities, *congruence* between client characteristics and program functions can be secured, while at the same time attending to the matter of program effectiveness and efficiency. Particularly, the different coverage modalities allow the minimizing of professional contact with those clients best able to sustain themselves, and allow greater guidance and decision assistance to those who are, for some reason or another, impaired in their judgment or deprived in their experience—irrespective of the type of social service being offered. The guiding ambition, of course, is to design the different coverage modalities such that, in aggregate, a given level of effectiveness can be secured at the least associated expenditure of administrative overhead. This means that the social service manager should always be looking to give remedial or compensatory face-to-face support where it is needed, but also seeking always to mechanize the delivery of services to that set of clients for whom interjection of advice or consultation would be deemed offensive or gratuitous. Thus, the coverage assignment process attempts to symmetrize effectiveness of service across a distinctly asymmetrical population. This, as will be seen in Chapter 4, is no mean trick.

There is yet another aspect to the matter of coverage modalities. This concerns itself with making distinctions about the different *periods of coverage* which might be employed. That is, once the client population has been subdivided in terms of sociobehavioral and conditional criteria, it will become apparent that concern with different clients will extend over different intervals. That is, some clients will be expected to fall within the purview of the system for a very long period of time, having no effective probability of becoming self-sufficient. These *chronic* cases would then be given an open enrollment period and should not be required to requalify except at very infrequent intervals. By the same token, some individuals would fall under the aegis of the system for only a limited period of time (these being, for example, individuals who are only temporarily *displaced*). The expectation for such individuals is that they will quickly become self-sufficient in certain relevant respects and therefore quickly become unqualified for the service the program provides. The period or frequency of requalification for this group should be rather high. It thus becomes important to make a three-dimensional correlation, one that assigns a distinct *coverage modality* to each distinct *client subset,* and which also establishes a potentially unique *qualification interval* for each client category. In this way, administrative overhead may again be reduced to an effective minimum by being able to restrict reprocessing according to the schedule at which individual client classes are expected to become unqualified. In short, transitory, displaced individuals will be reprocessed more frequently than chronic or secularly dependent clients. Always being looked for are system distinctions which simultaneously improve the probability of system effectiveness, decrease expected transaction (administrative) costs, and serve the interest of client dignity. Again, however, discussion of the methods by which this may be approached will be saved until the next chapter.

Referral Logic

The referral function of the typical social service delivery system has two aspects. First, it serves to direct individual clients to those providers ostensibly qualified to serve their needs. Second, the referral subsystem serves the integrative function spoken of earlier: the task of finding a path for the client to all social services for which he is presumed qualified.

In this second aspect, the referral function coordinates with the access function earlier discussed. It simply serves to maintain formal relations between different social service programs, such that a client need in effect contact only one social service program to have the resources of the others made available to him. In a properly organized, integrated social service system, this cross-referencing and transfer function might be served by a single front-end office designed to promote interchange as its major function. But the idea of a decentralized transfer mechanism is also attractive. The mechanics of cross-transfer are simple. First, it is necessary (or at least useful, given the criterion of administrative efficiency) to develop what are called *cluster functions.* In the simplest terms, cluster functions, pertinent to a particular community or domain, specify those casualty conditions which tend to appear in concert. The cluster function thus gives explicit substance to the proposition mentioned earlier: that a casualty on one dimension will often tend to be a casualty on other dimensions as well. Therefore, the cluster functions try to reflect the most likely *correlations* between various casualty conditions. Initially, these cluster functions may be a matter of judgment or opinion; however, as time passes, the initially subjective functions may gradually be transformed into empirically predicated ones, reflecting the actual (historical) clusteration of conditions which appeared in the community or domain of interest.

Once these cluster functions have been formulated, they will be used in the following ways: (a) those various social service programs which are expected to have shared clients—that is, those programs treating conditions that are expected to be significantly correlated—will be identified; (b) the agents of these correlated programs will then be supplied with the enrollment forms—and qualification criteria—for their sister services; (c) transferring a client from one program to another thus becomes a matter of administrative mechanics, and hence the rationale behind the transfer access modality earlier discussed; (d) in this way, the casual, informal networks that serve to move a client among different social service programs now becomes formalized, with the expectation that the bureaucratic demands on the client are reduced. For one agency with whom the client is in contact—equipped with an updated register of other social service programs and linked through the transfer mechanism—can affect the direction of the client to many different services without his having to initiate the contact or research the schedule of community services. In general, the cost of providing these transfer functions will probably be offset by the lowered enlistment demands on each of the individual programs. More research needs to be done on transfer logic, but it promises to be an effective mechanism for realizing administrative efficiencies, and is certainly an innovation designed to further client dignity.

The transfer aspect of the referral function is relatively straightforward and, with the exception of developing the cluster functions (which restrict the number of linkages which are pre-prepared or programmed to those most likely to be exercised most often), requires little technical skill. Not so with the matter of referring clients to providers, per se. Indeed, the matter of generating provider referrals is the most technically complicated aspect of social service system management, and one to which much attention will be devoted to in later sections. Here, however, briefly outlined is the general logic.

Initially, the referral function may involve very different things for different social service programs. For example, the concept of "provider" may have no real meaning to strict maintenance programs (those agencies which exist merely to transfer resources). Because (as already suggested) strict maintenance programs are something of an affront to the basic integrity of the social service sector, their characteristics will not be amplified here. Rather, it is the social service program—qua intelligent broker—which is of most concern. The broker reference is deliberate, for a broker serves not only as a fiscal conduit, but also has a professional responsibility to his client. This always implies that the broker is prepared to follow up any resource commitments with advice, consultation and some sort of audit. The implicit requirement for the proper broker is that he put his client in touch with that provider which appears to be able to deliver the best service. But, remember, that the broker also serves as the agent for the interests of the resource system (specifically, the taxpayers whose resources are preempted to the social service system). Therefore, his task is always to balance the interests of the resource system against the interests of the casualty system. In short, he pursues his client's welfare only within certain economic (efficiency) constraints. We are thus back to our original proposition about the managerial demands on the social service program: to strike a best (or approximately optimal) balance between effectiveness and efficiency. It is this proposition which sets the conditions under which the normative referral mechanism will operate.

First, there must be a strategic or policy-level decision about the critical factor: will the social service program provide its own services to the client, or will use be made of mainstream (private or quasi-public) providers? In the domain of medical care, for example, the question is between the establishment of public facilities dedicated exclusively to social casualties (e.g., county hospitals, public clinics staffed exclusively by government doctors) and sending clients to private physicians and private hospitals. The same sort of strategic issue emerges in other areas as well. For example, is it better to fund a separate staff of full-time public defenders, or to route indigent clients to mainstream attorneys? The relative cost-effectiveness of these two different strategies has really not been adequately researched (which is perhaps the surest sign that social service management is still in its infancy). Yet the decision about the use of dedicated versus mainstream practitioners must be made, though often it is taken by default.

What would the procedure be for making a more or less rational decision about whether to establish a public provider, or to use mainstream providers? The obvious answer is that an action-research program of some sort should be devised, such that the cost-effectiveness of the two modalities could be calculated in the context of actual operations. But there are also some abstract constraints which must be considered. For example, the dictate of client dignity might suggest that mainstream providers be employed to avoid the stigma of the client's being treated in a public facility (or to minimize the appearance of the client taking handouts). Moreover, considering that some areas have considerably different professional-client ratios (for example, different distributions of lawyers and doctors with respect to community populations), local factors have to be considered. Out on some remote Indian reservation, for instance, there may be no choice but to establish a public clinic or a dedicated, government-supported legal staff or educational institution. On the other hand, in most urban areas, mainstream care becomes feasible on almost any dimension. Mainstream care, however, is not feasible for all clients. For example, severely impaired individuals, violent or asocial persons, might have difficulty finding a provider willing to service them. Therefore, even in well-staffed urban areas, there might often be a necessity for having at hand some sort of public provider as a last resort.

As for the basic decision about public versus mainstream care for almost any type of casualty condition, basic economics probably provides a general answer. In specific, any public facility will have the same overhead as a private provider (rent, utilities, administrative overhead, etc.). Even when a private provider operates under prestige conditions (with a fancy office and modern equipment or technology) and the public provider locates in a poor neighborhood and offers a less dramatic technology, the differences between the fixed costs appear to be rather insignificant. Moreover, when considering the internal pricing system that would have to be imposed on government-employed professionals (including civil service benefits, personnel overhead, vacation pay, etc.), the charges imposed by private physicians, lawyers, etc., appear not to be excessive and may often represent a more economical investment.* When considering that the overhead for mainstream providers is distributed among a large population of clients, only some of which will be public charges, then *the aggregate economics generally favors the use of mainstream providers wherever possible.* The qualification is, however, simply this: that the social service agency be prepared to effectively audit the mainstream providers such that excesses and abuses are minimized. In short, it is only when there is essentially no rational audit function—and essentially sophomoric managerial technology—that there comes the kind of horror stories that emerge from the Medicaid-Medicare system. As many local health care officers are quick to testify, many of these abuses and shoddy practitioners would simply not be tolerated were decisions about providers left to local authorities rather than to a far-removed, mechanical bureaucracy.

From these general arguments come the demands that any rational referral system must be prepared to meet when the

*As will be pointed out in the next chapter, it is possible to develop reasonably accurate cost-effectiveness indices for providers without a priori distinguishing between the public agent and the mainstream professional. An objective (rationalizing) referral algorithm would thus distribute contacts as variable criteria dictate, and this saves us from making generalizations about classes of providers. In any case, some considerable research should be directed at resolving the issue about the comparative cost-effectiveness of public versus mainstream providers; the popular assumption that public providers represent a generally more economical investment is quite possibly erroneous, and perhaps dramatically so.

mainstream modality is to be implemented to any degree. As with the other social service functions discussed, referral technology will not be introduced in detail until a later chapter. But the generic logic should be clear even at this point. A proper referral mechanism must do the following:

1. Maintain an *audit index* which reflects the quality of the various providers available to deliver certain services. Thus, providers would be placed into various categories with each category suggesting a different expectation about the effectiveness of that set of providers with respect to the treatments presumed required.

2. Maintain a fiscal history on providers, such that the economy of using certain providers—given specific quality indices—is available before making a referral.

3. Maintain a distributive profile, such that referrals are spread throughout a large number of mainstream providers, rather than allowing referrals to cluster in a few areas or among a few practitioners (which would abort the exploitative providers such as those that have established Medicaid mills, etc.).

In all cases, then, a proper referral mechanism presupposes the existence of a *provider match algorithm* of some sort, such that referrals are made in a way which consistently attempts to realize a most favorable cost-benefit (cost-effectiveness) posture for the system in aggregate.

The central problem, of course, is that many professions either resent attempts at quality assessment or appear to be immune to the imposition of cost-effectiveness indices. As for the former, any provider who resists the attempt to be assessed may simply not be the right provider for any casualty client. Of course, quality indices should be kept very quiet, and the assessment process itself be made as invisible as possible, and where practicable linked with audit procedures supported by professional associations themselves.* As for the feasibility of imposing quality indices, modern management technology (and the judicious and ingenious use of surrogates and statistical processes) means that virtually no profession is immune to accountability. Therefore, there is little merit to the proposition that effective-

*As with the P.S.R.O. system sponsored by the medical profession.

ness or quality indices are inaccessible. Indeed, as will be seen in a later chapter, the HCDS system sets out a simple procedure for making such quality assessments for medical providers, and this technique would also be available for application to many other professional groupings.

In regard to the matter of possible professional recrimination which might occur were certain providers passed over because of unfavorable (system generated) cost-effectiveness indices, it should be noted that a proper referral mechanism will produce an *objective* basis for referrals. In fact, the type of assessment technique employed in the HCDS system would presumably be sufficient evidence against prejudice or caprice were the system to be sued by a provider for ignoring him. By the same token, the objective referral mechanism would also obviate the situation where a social service functionary routes clients to a single or a few favored providers because of some subjective preference or reward. In short, so long as the method is demonstrably objective, a majority of mainstream providers will not resent its use or development, and the system manager may avoid legal culpability for their referral decisions. Finally, it should be noted that an intelligent referral mechanism is the only way of ensuring both client and community (fiscal) interests. Public providers—dedicated to serving some indigent population alone—may tend to be inefficient and therefore a source of economic misallocation; where there is no control over private providers, then ineffectiveness and abuses may emerge with great frequency. Moreover, as often suggested, client dignity (with few exceptions) is usually *not* best served by leaving the casualty on his own, bereft of the advice and direction he might require.

At any rate, with this brief introduction to referral logic (and with the more detailed and technical discussions which will follow later), we may proceed to a very short discussion of the remaining functions of the typical social service delivery system.

The Clerical and Fiscal Logics

As may be seen from a final glance at Chart 3.1, the remaining system functions are: client tracking and record keeping, provider reimbursement, internal (fiscal) management, and report generation. The basic tasks associated with these functions will be detailed in chapters 4 and 5, where fiscal and data processing (clerical) procedures are discussed, respectively. Here are just a few preliminary remarks.

Initially, a client's history in the system is kept track of for the following fairly obvious reasons: (a) to see that he actually receives the treatments for which he was scheduled or recommended; this, then, seeks to control the effectiveness of the program with respect to individual clients; (b) to see that no individual client abuses the privileges of the system or attempts to exceed the coverages for which he was qualified; (c) to ensure against providers attempting to bill for services that were not actually performed; (d) to control the expected size (probabilistically) of the client population by being able to predict attrition of existing enrollees, etc. Again, these are mainly clerical functions, and the major concern of Chapter 5 will be to show how they may be performed most economically within the context of a real-world social service delivery system. Here it is enough to simply suggest that these tasks do indeed have to be performed.

The provider reimbursement mechanism is also relatively straightforward, yet Chapter 5 will elaborate on its properties considerably. Particularly, the reimbursement system serves as the *transaction* mechanism for the system, and thus becomes intelligible in terms of normal accounting procedures (e.g., the management of accounts receivable and payable). In practice, there must also be a provision available for auditing requests for reimbursement and for making reimbursement contingent upon certain basic quality considerations. Also, the information generated by the history of reimbursement requests for a provider gives the information needed to assess the relative cost of that provider, and thus yields one aspect of the information required to generate the cost-effectiveness indices used by the referral subsystem. Finally, as will be seen in Chapter 5, there are certain organizational tricks which can allow one to effectively minimize the transaction costs associated with social service operations.

In the matter of fiscal control, as already suggested, the two most important characteristics are: (a) the rate of fiscal expenditures (resource commitments) must be linked back to the threshold qualification requirements in the access subsystem, such that the system neither overspend nor underspend during a fiscal period (in Chapter 5 are illustrated both the theoretical and practical logic by which rates of expenditure may be rationalized over a given operating period); (b) the fiscal control system also serves as a major source of information for system planning, particularly with respect to the resource-population (per capita resource) ratios which may be expected or which *should* be sought.

The last of the subsystems—the final component of Chart 3.1—is the reporting function. This subsystem has basically three functions: (a) to meet the reporting requirements imposed by external funding authorities (such as Medicaid, etc.); (b) to produce summary operational histories which may be used by the system manager as current inputs into the fiscal control subsystem, and as a continuing source of information about possible areas of abuse, ineffectiveness or inefficiency within the system's current operating posture; (c) to provide the data necessary for strategic programming, such that the goals and objectives and basic characteristics of the social service program may be adjusted to meet projected changes in the environment or client population. Again, all of these functions will require some technical explanation which will occupy us later; in the context of outlining the data processing and reporting logic built into the HCDS system, illustrated will be what reports are most likely to be useful, and something of their managerial and programming implications will be suggested.

FOCI OF SYSTEM PERFORMANCE

From what has just been done, it should be apparent that each of the major components or subsystems of a social service delivery system would have its own focus of performance. The aggregate performance of the system—say, some overall, encompassing *cost-effectiveness* index—would then be built up from a complicated interconnection of the several parochial performance indices. This matter of adjudging or measuring cost-effectiveness in the social service sector is, of course, one fraught with difficulties, dispute and equivocation. In the space available here it is impossible to give much more than a simplified, synoptic view of the process. But even so, it will be clear that the contents of Table 3.3 have some interest for us.

Initially, *elasticity* of access measures the cost associated with the effort to enroll—and to identify, inform, motivate, etc.

TABLE 3.3 / Dimensions of Performance Analysis

Subsystem	Performance Focus
1. Access	*Elasticity:* the proportion of the casualty universe accessed into the system at a given rate of expenditure
2. Qualification	*Consonance:* the maintenance of a favorable balance between client population and the resource base
3. Coverage Assignment	*Congruence:* development of favorable cross-correlations between client-conditional properties and coverage (delivery) modalities
4. Referral	*Optimality:* the delivery of an adequate level of service at tolerable expenditure . . . the affecting of consistently favorable resolutions of the tradeoff between service and economy
5. Fiscal Control	*Integrity:* equitable distribution and exhaustion of resources across temporal or capitation intervals
6. Tracking/ Reporting	*Leverage:* the maximization of the utility (usage) of each unit of information generated by the system (or each quantum of data stored and manipulated)

—clients in the social service program.* In absolute terms, elasticity becomes a per capita index reflecting the average cost of accessing a social casualty. In relative terms, elasticity would be deflated by a proportion . . . the proportion of the total casualty universe actually brought into the program. Therefore, access effectiveness implies that the access function was able to reach some number of clients relative to the total expected (estimated) casualty population, while the efficiency measure simply measures the per capita cost associated with the actual number of clients enrolled.† The sister function, qualification, is a complicated function; hence, *consonance* is a complex index. The qualification process should not be a passive function. Rather, the variability of the qualification threshold is one of the key management instruments available to the social service administrator. Consonance, in this respect, thus implies that the qualification threshold (and hence the client population) was raised or lowered

*Note that the term elasticity is used here to measure performance in the access or outreach functions in light of the more widely used concept of promotional elasticity . . . a measure of the sensitivity of demand for goods or services as a function of advertising expenditures.
†A better denominator, in most cases, is the number of clients who actually sought enrollment, as a system may be expected to access some individuals who are, in fact, not eligible given current qualification criteria.

with respect to interval-in-time resource base values in an effort to maintain a favorable (or at least adequate) per capita resource level for the system. That is, qualification exercises should see to it that the number of clients enrolled in the system does not unduly erode the per capita resource base and thereby reduce the rudimentary effectiveness of the system by diluting level of service. By the same token, consonance would also imply that the per capita resource level is not allowed to rise so high that decreasing returns to scale set in (or that the program concentrates resources in certain client strata at the expense of the broader casualty population). Therefore, operationally, consonance in the qualification function implies a constant coordination between client population and resource base and is one of the critical managerial tasks the proper social service administrator must perform.

Moving on, *congruence* with respect to the coverage assignment function implies that the system develop and maintain different delivery (coverage) modalities, and that clients are assigned to one or another of these modalities through a rationalized process. Particularly, as has been suggested, (and as will be considerably amplified in the next chapter), clients will tend to differ both in terms of their sociobehavioral and conditional (casualty) attributes; therefore, no single delivery modality can be effectively optimal with respect to all clients. The coverage assignment task thus seeks to make sure that clients are assigned to that particular modality most congruent with their individual properties. The performance focus for the referral subsystem, *optimality,* should always be interpreted as an approximate criterion. In effect, it asks that each time a client is referred for a treatment of some casualty condition, the provider selected to deliver the treatment be the one with the most favorable cost-effectiveness index. In such a way, an adequate quality of service is ensured while, also protecting the economy of the system. Little more shall be said about this, except to mention that the logic underlying the attempt to make optimal referrals is explained in some detail in the last section of Chapter 4. (where the matter of developing cost-effectiveness indices for individual providers is also treated). At any rate, it is hoped that the reader will soon become aware that the matter of referral optimality and the congruence of the coverage assignment process is the very heart of the rationalized social service system and the determinant that most directly affects the overall cost-benefit or cost-effectiveness ratio a system will obtain.

As for the matter of *integrity* in the fiscal control (and reimbursement) subsystem, this involves the way in which the funding bases or financial commitments of the various resource providers are handled. In general, the system manager should completely use the resources provided him by the various input sources, and thus exhaust budgets within the temporal or capitation limits the payors specify.* That is, some budgets are given for a fiscal year; other budgets are based on a formula which projects expenses across a definite client population (or provides for a certain definite number of treatments to be provided). In either case, the system manager must seek to manage the funds so that they are distributed equitably across the client population, trying to avoid instances where overexpenditure on some clients reduces service available to others, as well as instances where the budget is underspent because of excessive conservatism or irrational fiscal constraints. There are many qualifications surrounding this problem of maintaining integrity, most of which will be discussed at some length in the beginning of Chapter 5.

The last of the performance foci, information *leverage,* is simple to explain in concept, but difficult to realize in practice. Specifically, we maximize information leverage when we make each unit of information we generate serve the largest number of functions. That is, each unit of data the system produces or manipulates should first of all have definite decision (or reporting) implications, such that no records are kept which do not have a unique utility. Secondly, with some rare exceptions, the cause of information leverage is best served

*This is largely a matter of convention and certainly not a normative suggestion. Obviously, were social service management built around the concepts of minimizing opportunity costs in all investments—and allocating resources on the basis of comparative cost-effectiveness indices—then integrity would dictate that managers forward excess funds for redistribution. But the formula-budgeting system employed as part of the categorical funding mechanism does not explicitly sanction such an exercise, nor does the long interval between preparation of new program or zero-based budgets facilitate intraperiod transfers. However, in a localized, integrated social service program, the fiscal authority would not be stopped from transferring funds among categories, either within or between fiscal periods. Therefore, from the normative perspective being urged, all expenditures should be rationalized and not just geared to fixed ceilings. From the perspective of the practicing manager forced to cope with the categorical funding mechanism, the most attractive strategy is, indeed, to exhaust the budget allocated him or face budget reductions for the following year.

when each datum appears only once in the information base.* Finally, individual forms and files should be structured so that reporting requirements are the key determinants of the information generation functions, and such that reports may be generated *on demand* rather than on the basis of irrational requirements (which usually emerge when someone decides to collect data first and looks for applications only after the fact). These points and the practical aspects of information leverage in general are given more detailed discussion in the first and last sections of Chapter 5.

In the chapters that follow, there will be an opportunity to show how a real-world system was designed and implemented with these performance criteria in mind. Moreover, as we now begin to analyze the components of the HCDS prototype, we shall lend substance and operationality to virtually all of the abstract, generalized points made in these first three chapters. In short, we are now going to shift analytical gears, turning from the general to the specific, and from the speculative to the pragmatic. More important, perhaps, we can now leave our concern with what *should* be done and spend the remainder of this volume dealing with things that *can* be done.

NOTES AND REFERENCES

[1] As a good introduction to the way in which modern market research attempts to control or contain confoundations, see Massey, et al., *Stochastic Models of Buying Behavior* (Cambridge, Mass.: MIT Press, 1970). Most of the discussions in this interesting volume are extensions of Markov logic; some readers might find the Markov formulations useful as a tool for predicting casualty populations and for computing casualty correlations.

[2] For more on this, see Chapter 1 of John W. Sutherland, *Administrative Decision Making: Extending the Bounds of Rationality* (New York: Van Nostrand Reinhold, 1977). For an excellent analysis of the need for combining theory and praxis in science in general, see James B. Conant, *Modern Science and Modern Man* (New York: Columbia University Press, 1952).

[3] For a brief note on what heuristic organizational strategies might look

*This rule usually refers to information systems which employ what is called a random access capability. Where sequential access is dictated because of mechanical or technological limitations, it is sometimes useful to develop data redundancies, simply because retrieval costs may be very high for exclusive data.

like, see John W. Sutherland, "Attacking Organizational Complexity," *Fields within Fields . . . within Fields* (vol. 11, Spring 1974).

[4] To see what alternatives to hierarchical, bureaucratic structures might look like, see Herbst, *Alternatives to Hierarchies* (Mennen Asten, Netherlands; 1976); or Chapter 8 of Alvin Toffler's *Future Shock* (New York: Random House, 1974); or Warren Bennis' "Beyond Bureaucracy," *Transaction* (July–August, 1965), or his *Changing Organizations* (New York: McGraw-Hill, 1966).

[5] For notes on professionals in formal organizational settings, see the appropriate sections of Blau and Scott's *Formal Organizations* (San Francisco: Chandler Publishing Company, 1962), or John W. Sutherland, "Towards an Array of Organizational Control Modalities," *Human Relations* (vol. 27, no. 2, 1974).

[6] Cf., *Experimenting with Organizational Life,* ed. Alfred Clark, (New York: Plenum Press, 1976).

4

THE CLIENT DIMENSION

INTRODUCTION / We have now reached the point where we can talk about the details and mechanics of social service delivery systems. This means that we are going to stop arguing in generalities and start looking at a particular system in specific detail. Different aspects of any social service delivery system tend to become apparent from various perspectives. In the two remaining chapters of this volume, we shall be adopting two dissimilar viewpoints and seeing the HCDS prototype system from two rather different angles. In this chapter, we shall be looking at the HCDS system from the client's standpoint, and thus be concerned primarily with matters of access, qualification, screening, referral, etc., all the things that are apparent to the individual being processed by the system. In the fifth and final chapter, we shall look at the HCDS system on the *dollar* and *data* dimensions from the perspective of the system manager and data processing functionaries.

AN OVERVIEW OF CLIENT LOGIC

Certain aspects of any social service system tend to be of different interest to the *client* himself; others are the primary concern of technicians or administrators and may remain beneath the client's awareness (and properly so). Indeed, it is generally possible to develop three very different logics for any delivery system. The first of these logics would be the way in which an individual client is processed through the system. A second perspective would be the *dollar dimension,* where the concern is primarily with the integrity of the resource base and with matters of economy and efficiency. Finally, to maintain the coherence or organizational integrity of a system, the data or informational dimension must be considered. The *data dimension* has no real life of its own (except perhaps for the computer programmers or system analysts), but rather serves as a subordinate to the client and dollar dimensions.

At any rate, from the client's standpoint, the concern is with the system components that are directly or tangentially in the path of the individual. The emphasis from this perspective is twofold: first, to see that the services provided the client are effective, given the objectives and interests of the program; second, to see that the criterion of client dignity is served. In these respects, see Figure 4.1. In great abbreviation, these are the provisions (taken from the client's perspective) that one would expect to find in virtually any properly designed social service delivery system. The client, initially, gains access to the program

FIGURE 4.1 / A generalized client logic

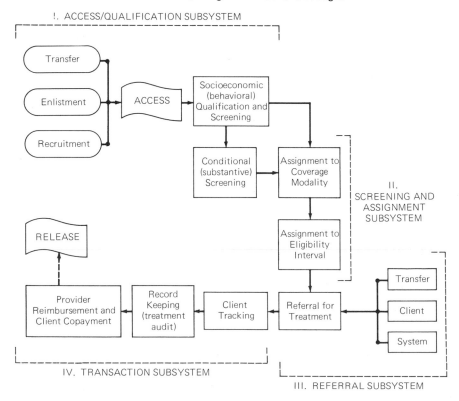

through one of several different mechanisms, is screened both for his socioeconomic qualifications and his conditional (e.g., medical) attributes, and then assigned to one or another of the different coverage modalities appropriate to his socioeconomic and conditional properties. The coverage (or delivery) modality to which he is assigned also, as a rule, will determine the eligibility interval assigned him (this being the time allowed between requalifications).

Once the client has been assigned a coverage and eligibility designation, he is then eligible to receive specific treatments. Every social service system will have some sort of rationalized referral mechanism which should be able to point the client to the most cost-effective provider (given various criteria for selection). Once a referral has been made, the client is then "tracked" to see that his treatments have indeed been received, his records are updated and, if he is a copayor, he is billed for his portion of the costs (or perhaps there is a simple deductible system at work). At any rate, the remainder of this chapter will be devoted to an analysis of the way in which these functions are performed in a specific system, the HCDS prototype.

THE MECHANICS OF ACCESS AND QUALIFICATION

The front end of any social service program will generally consist of its access and qualification mechanisms. There are two reasons for referring to the access and qualification mechanisms as the system's front end. First, there are the mechanisms with which any potential social service client first comes into contact and which either channel him into the system or deny him enrollment. Second, the access and qualification mechanisms are the system components most responsive to *policy* provisions. That is, the type of access employed—and the nature of the qualifications which are posed for system coverage—are determined as a matter of basic policy formulation, and thus represent the primary linkage between the prevailing political system and the casualty system. Qualification criteria determine who will be enrolled in the system, while access policies generally determine the posture that the system will show to the community at large. In short, it is through the access and qualification mechanisms that any social service program gains its political significance. It is thus important that any delivery system be responsive to changing political interests, and this means that the access and qualification mechanisms should be flexible. For as things now stand, social service programs must be viewed as an extension of the political system, and not as something "above" politics.

As suggested in the third chapter, the access and qualification mechanisms must also be responsive to prevailing economic conditions; of course, the very existence of social service programs implies a certain structure of prevailing social sympathy. Thus, the context in which any social service system is forced to operate is a complex one. It exists, in the strictest sense, in the constantly changing nexus of political, economic and social pressures. As previously suggested, the aggregate social service sector is dependent upon political functionaries taking a reading of prevailing social opinion and then funding programs to the extent that opinion recommends. So too with individual social service programs. Should prevailing public opinion shift toward individualism and away from a paternal attitude toward indigents or social casualties, the political capital of social service systems is diluted. Expectedly, the funding base will be reduced, and the number and scope of social service programs will decline. On the other hand, should a significant enough proportion of the general population find some personal advantage in social service coverage, then political interest may be expected to accelerate and the social service transfer funding would increase. The economic variables play an obvious role. In periods of recession, for example, two conflicting tendencies emerge. The resource base (the noncasualty population) becomes increasingly irritated at the erosion of disposable personal income due to the increase of resources preempted by the social service sector. But, on the other hand, the casualty population tends to increase with unemployment and the decline in capital investment, etc. In times of economic boom, increased general affluence tends to take the sting out of transfer payments for the taxpayer, and also leads to a decrease in demand for social services due to the decline in the indigent (casualty) population.

It is beyond the scope of this book to try to show all the various permutations of social, political and economic conditions and their implications for the social service sector. But it is important to recognize that the social service sector is caught up in a constantly complex *dialectic* between forces acting to protect and extend it, and forces acting to restrict and dismantle it.

Again, this dialectic recommends that the social service program maintain a *flexible front end* and avoid casting either its access or qualification mechanics in concrete. Where this has been done, even in the relatively stable socioeconomic context enjoyed in recent decades (though, to be sure, there have been some perturbations), dysfunctions have been spawned. For example, the Social Security system is approaching abject insolvency because of its inflexible support commitments in the face of a greater-than-projected client population; the social service programs that have been especially victimized by the fiscal despair of New York City have been mainly those with inflexible qualification criteria and with access mechanisms that are not controllable.

Rationalizing Access

A look at the access and qualification mechanisms built into the HCDS prototype lends some substance to these arguments. In this regard, consider Chart 4.1 which is the first of a new set of logic charts, this set being the *reference* master logic. This and the four other reference charts (to follow in this and the next chapter) differ in two respects from the *normative* master logic charts introduced in Chapter 2. First, the reference charts respond to certain technical or contextual constraints that were ignored by the normative designers; in short, the reference logic takes explicit consideration of restrictions that were not known to the design team in the normative phase. Second, it must always be expected that the normative design will be altered in some respects when the designers come face-to-face with the realities of implementation. The reference logic, then, takes account of the fact that the designers will learn more about their project as time goes on. Thus, the reference logic charts differ in some respects from the normative charts. This is why the reference charts will be explained in great detail, and why the normative charts were introduced for illustrative purposes only.

At any rate, Chart 4.1 begins by noting the means by which clients gain access to the HCDS. Particularly, four different access modalities are available, thus giving the flexibility which was just argued for. The first of the access mechanisms, allowing some other social service agency to promote a prospective client, implements the *transfer modality* talked about in Chapter 3. The point to the transfer modality, again, is that casualty conditions

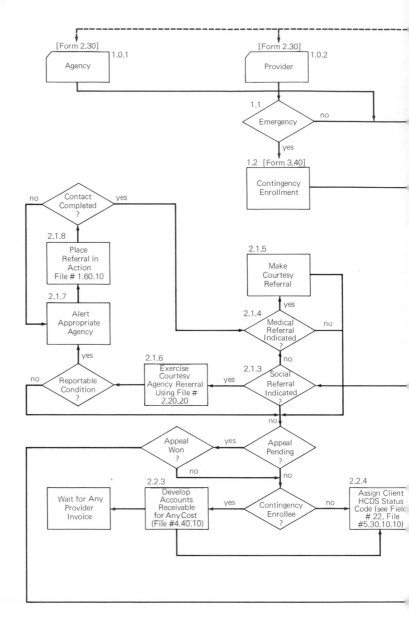

CHART 4.1 / Access and eligibility subsystem

in any community tend to be correlated or clustered. Given this realization, the mechanics of access via the transfer mechanism become fairly straightforward. First, an effort should be made to actually construct *casualty correlations* for a community. These

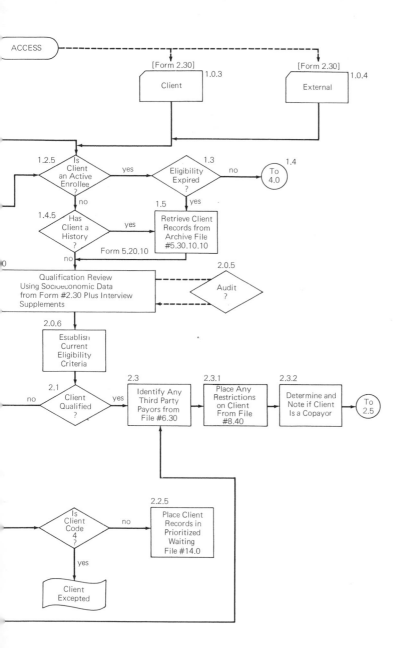

ACCESS

[Form 2.30] 1.0.3
Client

[Form 2.30] 1.0.4
External

1.2.5 Is Client an Active Enrollee ? —yes→ 1.3 Eligibility Expired ? —no→ 1.4 To 4.0

no ↓ | yes ↓

1.4.5 Has Client a History ? —yes→ 1.5 Retrieve Client Records from Archive File #5.30.10.10

Form 5.20.10

no ↓

Qualification Review Using Socioeconomic Data from Form #2.30 Plus Interview Supplements - - - 2.0.5 Audit ?

2.0.6 Establish Current Eligibility Criteria

no ← 2.1 Client Qualified ? —yes→ 2.3 Identify Any Third Party Payors from File #6.30 → 2.3.1 Place Any Restrictions on Client From File #8.40 → 2.3.2 Determine and Note if Client Is a Copayor → To 2.5

Is Client Code 4 ? —no→ 2.2.5 Place Client Records in Prioritized Waiting File #14.0

yes ↓

Client Excepted

casualty correlations will attempt to show what conditions—covered by different social service agencies—will tend to appear in concert with respect given individuals. It is expected, for example, that many social casualties will be simultaneously the concern of some educational development agency, some health care program, some job-retraining (or placement) bureau, etc. To the extent that such correlations exist, we then have the basis for implementing the cross-transfer mechanism spoken of in the last chapter. Particularly, the casualty correlations will identify those agencies that will probably *share* clients in common, and which should therefore keep each other supplied with enrollment forms and updated with respect to current enrollment (or qualification) criteria. In this way, a client may be enrolled from any single source agency in all the programs for which he is eligible without having to deal with several different bureaucracies. Moreover, the enrollment and access overhead costs are thus distributed throughout a large number of sister agencies, and thus become intelligible as the expenses of de facto integration. This cost of integrating a local social service system is minimized to the extent that *only* significantly correlated agencies are asked to maintain cross-contact and to provide for transfer access. Agencies which are unlikely to share clients are not expected to support transfer linkage. In short, the interagency network is *rationalized* in that cross-contact provisions are dictated by calculated probabilities of interchange. As a final note about the transfer access mechanism, sister agencies should be expected to supply each other with enrollment forms on the basis of the *strength* of the correlation functions, so even the production of documents may be disciplined by proper analytical techniques.

Moving on to box 1.0.2, note that *providers* themselves may be authorized to access clients for the HCDS system. The rationale here is simple. Mainstream professionals (doctors, lawyers, etc.) will often be approached by clients seeking a service; in some cases, those clients will be eligible for support from a social service program. But the client may often be unaware of the program's provisions or even its existence. To the extent that various mainstream providers are made aware of the type of coverage which a program provides and kept informed of current qualification criteria, the mainstream providers themselves become part of the program's access system. Say a patient comes to a physician with a complaint. It may become apparent to the physician that the patient needs treatment but cannot afford it. If the physician is aware of the HCDS, he may then actually initiate a request for that patient's enrollment (and may, under the provisions of the HCDS, actually designate himself as the preferred provider, subject to qualifications discussed in the last section of this chapter).

Associated with the mechanism permitting provider-driven access, there is a question box (1.1) that asks if the provider's contact with a patient was in the nature of an *emergency*. If so, then follow the logic to box 1.2, where the provider's patient is given what is termed *contingency enrollment* in the HCDS. This is an interesting provision, and one that has considerable potential for ingratiating a health care system with the general community population (and not simply with the medically indigent). In the simplest terms, the contingency enrollment provision implies the following: any provider (clinic, physician, hospital, etc.) faced with an emergency situation may offer the necessary treatments with the assurance that the HCDS will be responsible for reimbursement should the patient be unable (or unwilling) to pay for the services rendered. In short, a provider operating anywhere in the community served by the HCDS knows that when he treats an emergency case, he will be properly and promptly compensated. Moreover, where resources or circumstances permit, it may be possible to indemnify providers who offer emergency treatment against malpractice suits, this protection being offered through some kind of block insurance policy. The purpose of the contingency enrollment scheme is thus to make sure that all citizens in a community, in case of emergency, are given treatment at the most proximate provider, even when they cannot demonstrate financial responsibility by producing an insurance card or a checkbook. In short, any citizen suffering an emergency is now taken to the nearest provider, and not transported to some public hospital that is perhaps further away but less restrictive in providing treatment. The same logic holds true for clinics or private physicians as well as hospitals; any demonstrable emergency implies contingency enrollment, and contingency enrollment means that no provider need be concerned about supplying treatment even to patients he has never seen before and probably will never see again. For many citizens this provision may be of some comfort, especially in those communities where emergency cases are all transferred to some public provider unless the patient is alert enough to indicate a preference and meet financial criteria. It is thus that, with a bit of imagination, a social service program can show itself of benefit to the population as a whole and not simply to the casualty subpopulation.

In terms of mechanics, the contingency enrollment may work this way. Once an emergency is demonstrated, a provider is able to bill HCDS directly for any legitimate services. HCDS will then reimburse that provider. Should a patient ultimately prove to be financially responsible, then HCDS will seek to be reimbursed for its expenses. The provider does not, on his own, have to support any action against the patient. The HCDS thus absorbs both the risk of nonpayment and the cost of collection. In many cases, patients who do have third-party coverage or sufficient personal means may choose to repay the provider directly, in which case the contingency enrollment option is not implemented. In any case, to the extent that a program's resource base permits, the contingency enrollment strategy should be carefully considered, either as outlined here or in some local variant.

The third means of entering the HCDS system is self-access, which is the counterpart of the *enlistment* modality discussed in Chapter 3. The question that arises about enlistment is always this: should just a single point of access be maintained, or should there be decentralized access points? Then there is a subordinate issue: If several access points are elected, where are they to be placed? It is interesting to note that these are essentially the same questions we would ask were we thinking not just about providing access, but about providing services themselves (were something other than mainstream providers used). For example, were a health care system to develop public clinics to deliver its services, the matter of location becomes analytically the same as the matter of access. Certain management science instruments can help with such decisions.

First of all, it should be clear that we need to develop some sort of algorithm that describes the way in which the casualties would be distributed across the community a social service program is to serve. As a general rule, this would imply the development of *density distribution* functions. Consider, for example, that every community or domain in general can be mapped. The density distribution functions would be superimposed on this map and would have the effect of showing where potential social service clients are located. Moreover, these functions would indicate whether or not there are any significant *clusters* of clients, i.e., whether different geographic areas tend to have different concentrations of social casualties. Now, as a general rule, when the density distribution functions indicate that a program's potential client population is distributed symmetrically (stratified) across the domain of interest, such that no clusters result, then there is a strong argument for just a single access or service station. Or, where the domain is very large and the potential client population too massive to be accommodated at a single point, it would argue that access or service stations be located strictly with regard to geographic considerations, e.g., locate a station in each of the four quadrants of the domain or locate a station on either side of a major river or roadway.

In most communities, however, the casualty population will not be distributed symmetrically. Rather, certain definite clusters will appear. In smaller towns there may tend to be only one cluster, this probably being located in what is popularly referred to as the wrong side of the tracks. Therefore, the service or access station should be located with respect to the cluster and not in the center of the community, per se. In larger communities and in many rural areas (or where a program is to serve a region rather than a specific locality), there is often a more complicated situation: social casualties may be located in several different clusters, each of which may be separated by some distance from another. In such a case there is a somewhat more complicated location decision.

Initially, the decision is whether to try to get away with a single access/service station or develop multiple stations. If multiple locations are chosen, then the next decision is on their exact placement. The unfortunate aspect to such an analysis is this: the first issue cannot be settled without previously deciding the second. Therefore, a rather special type of analytical approach is required, and the helping tool here is *centrix* analysis. Though it is impossible here to go into the mathematics of the process, an intuitive feel for what it can do can be given. In this regard, consider Figure 4.2. These models assume that the domain of a social service program may be translated (through topological transforms)[1] into a circle, and the locations of casualties projected onto the perimeter of that circle. In part (a), given that the client population is distributed with more or less constant density across the domain, a precisely central location is determined. This means that the average client is as well served as possible, given the distribution.

But where the client population is significantly clustered, then the symmetrical solution proves suboptimal; some clusters would be distinctly ill served compared to others. Therefore,

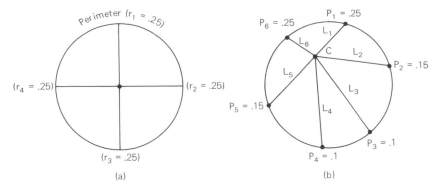

FIGURE 4.2 / (a) The symmetrical solution (assuming a random or symmetrical distribution of clients and/or casualty conditions across the domain of interest). (b) The centrix solution (assuming an asymmetrical or clustered distribution of clients and/or casualty conditions across the domain of interest).

some sort of centrix algorithm is called into play, as with part (b). Here the different clusters are all assigned density values reflecting the intensity of clients and/or conditions associated with specific locations (e.g., areas on a grid, subcensus cells or Cartesian coordinate plots). These density values (the p_i's) all sum to unity, and thus may be treated as probability density functions. Now, it is not strictly necessary to project the clusters onto the circumference of the circle ... this is just an illustrative artifice. It would be just as correct, and perhaps even more expedient, simply to evaluate each of the plots on a grid map. In either case, however, the distance from a cluster to the centrix (the location established for service or access) is a function of the reciprocal of the density function: $L_i = f[1/P_i]$. Thus, the factors of the location of clients and their intensity of client population across a domain can be used to develop a solution that is more responsive than the simple symmetrical solution. For the placement of the access/service center is now such that the distance from the center to the average client (or cluster) is effectively minimized across some complex domain. It is useful to conduct a centrix analysis using successively enlarged or reduced domains of interest and in such a way trying to determine the optimal number of locations as well as optimal placement. Again, there is no difficulty in using any type of domain function ... the configuration may be a rectangle, a triangle or virtually any of the more complex forms available; even existing census tracts can be used, providing that the rules of topological transformation are not transgressed. It is also possible to use qualitative inputs to the centrix analysis. In this case, instead of numerical density functions, simple ranking values like very high density, low density, etc. could be used.

In any locational analysis of the type pertinent to social service operations, there may be a centrix point determined for the community as a whole, or several different points determined on the basis of the successive enlargement and reductions of the centrix domains (as part of a gradual convergence on an optimal number of locations). At any rate, given the flexibility of the domains as defined within the centrix process, other management science instruments can be employed to gradually work toward an optimal access/service configuration,* as follows:

1. Initial development of a *queuing* model to suggest the rate at which clients might be expected to appear for service at the central station.[2] This would give an idea of the rate of service the station must be prepared to provide which can then be translated into cost implications.

2. Similar queuing models would be established to show the service rates required were several different locations to be employed, again translating these service rates into cost implications.

3. Next, would be the imposition of different tolerable waiting times on the single and multiple locations, and trying to isolate the best balance between service rates and the costs of maintaining various levels of service.

4. *Simulation* of the combined effectiveness alternatives (different waiting times and transportation demands, etc.) and cost alternatives (the functions suggesting the way costs rise with decreasing waiting times and/or providing multiple stations), and an attempt to see the most favorable tradeoff point.[3]

5. Then a simoptimization model[4] might be used to evaluate the effects (on cost and service) of expanding the cluster definitions ... of developing centrix locations that serve two or more of the original clusters as opposed to centrix loca-

*And whether the multiple stations should be staffed continuously or in some sort of rotation.

tions being dedicated to a single cluster or to community as a whole.

The net result should be an approximately optimal placement of access/service stations, a placement which best meets the competitive objectives of efficiency (cost minimization) and client service (and dignity). For example, a single location for an entire community may be so inconvenient for some clients that they simply do not appear; hence, the system's effectiveness is impaired. On the other hand, the frequency with which multiple stations are established by some social service programs (under the neighborhood strategy) means that overhead may be consistently higher than it should be. At any rate, it is well within the capabilities of properly trained management scientists to resolve such issues and thereby discourage suboptimal access/service schemes.

The fourth and last of the access mechanisms may be disposed of very quickly. This is the case where external sources pose clients for enrollment and, hence, becomes intelligible as the HCDS version of the recruitment modality introduced in Chapter 3. Without going into detail, the casualty correlations and density distribution functions just developed should be able to help decide who should be a recruiter and where they should be located. The casualty correlations suggest what type of recruiter should be sought, what functionaries are most likely to come into contact with clients. The density distribution functions locate clients and thus suggest where recruiters are likely to be most frequently sought out by clients. And, of course, the audit considerations discussed in the last chapter will help determine the precise recruiter population to be maintained, and eliminate those recruiter classes (or individuals) who tend to forward unqualified individuals for enrollment.

As a final note, there are two general aspects of the access function that should be mentioned. First, the overall objective for any social service access mechanism is to reach the largest possible client population for any given level of expenditure. Shortly, this assertion will be qualified by suggesting that the proportion of the casualty population actually accessed should be determined with reference to the system's ability to accommodate them at any point in time (relative to existing population-resource ratios). But the demand for an approximately optimal tradeoff between cost and effectiveness still pertains. For this reason, it is apparent that the access function should be *decentralized,* that external recruiters, cross-agency mechanisms and mainstream professionals (providers) be allowed to take as large a share of the access burden as they are willing. These mechanisms (as explained in Chapter 3) are less directly expensive than the enlistment mechanism from the standpoint of the social service program itself and therefore tend to favorably affect the cost-effectiveness index of the access function (i.e., they tend to lower the per capita access cost).

The second but related point is this: discussion of access logic helps to again point up the distinctions to be made between effectiveness and efficiency. Particularly, a program that uses only the self-access (enlistment) modality may get clients into the system, and therefore be considered absolutely effective to some degree. But the program may, in this regard, be inefficient, for it has elected the most expensive of the various access mechanisms and ignored those with an expectedly greater *elasticity of access* (greater leverage relative to cost). Therefore, it is no great thing to be effective, per se, if this effectiveness comes at the expense of efficiency. Now, from what we already know, we can see that any inefficiency always rebounds to the detriment of effectiveness. For example, when the access subsystem is inefficient, it wastes resources that might otherwise be available to provide services, per se. Because actual system effectiveness is always a function of the level of service a system can provide, any misallocation of resources in any component of the system (any instance of inefficiency) ultimately translates into an instance of ineffectiveness.

Now, these rather obvious points are mentioned for the following reason: it is commonly asserted that a system can be effective without troubling itself about the potential of management science or other formal analysis instruments. That is, the science of management is often thought to be something of a luxury. Since effectiveness is related to efficiency—and related inextricably—any opportunity to approach optimality that is foregone by social service administrators feeds back to the disservice of clients and erodes effectiveness. In short, except perhaps in the arenas of war, politics and love, effectiveness has no meaning unless it is qualified by some efficiency factor. Therefore, the type of access logic set out here cannot be considered a luxury or be ignored with impunity. Indeed, the only occasion on which any management science or formal analytical instrument should be ignored (or on which an opportunity to approach optimality should be foregone) is when it would cost more to employ (or realize) than it is really worth.[5] In such a case—and only here—can the administrator legitimately exercise his intuition.

Notes on the Qualification Process

Considerable time has been spent on the access process, relative to the space its functions occupied on Chart 4.1. But we may now move rapidly through an explanation of the remaining components of the rest of the access-qualification logic chart, beginning with box 1.2.5. Here, once a client has passed through one or another of the access channels, the question is asked whether or not the client is already an enrollee. If so (following the yes path), has his eligibility expired? If not, then simply forward the client to the referral system (see Chart 4.3). If, however, the client's eligibility has expired—or if he has previously been enrolled—retrieve his records from the archive and proceed with the central business at hand, qualification.

The qualification process itself is rather simple. It is an attempt to enduce information about the client's socioeconomic and casualty status in an effort to determine if he is indeed qualified to be enrolled in the program. As suggested in the previous chapter, there should also be some sort of *audit* scheme at work (box 2.0.5). This would, on the basis of some sampling scheme (probably associated with a stratified design), elect certain enrollment requests for further analysis and verification. It is desirable, of course, to keep the number of enrollment requests actually audited to some sort of effective minimum. For this reason, some working knowledge is needed of the access modalities or client subsets (population strata) which are most likely to forward unqualified candidates.

Aside from this, there are only two other points to be made about the qualification process, per se. First, it was suggested that qualification criteria should be variable with respect to the prevailing population-resource ratio of the social service program (or with respect to projections of client demand and available resources). Therefore, the qualification personnel should have some sort of algorithm at their disposal that suggests how qualification thresholds should vary with respect to population and resource parameters. This implies, in effect, that every social service program should have a *demand function*. The social service demand function would do the same thing as the demand functions developed by commercial concerns. It would suggest how effective demand for enrollment would change as the qualification thresholds are eased or made more restrictive.* That is, we would want to know how much or how little to alter the qualification threshold with respect to changing resource bases.[6] This demand function would be developed using the density distribution functions spoken of in the previous section and would therefore become an essential instrument of rational social service management.

The second aspect of the qualification process is somewhat more speculative but of equal importance. Particularly, we have to ask about the definition of the client population, and ask about it in broader terms than in the past. With the arguments about the necessity for providing a *national* health care program of some sort, the question of the client population becomes critical. As can be seen, the type of HCDS being developed here promises to be a considerably better system than one which simply gives everybody an insurance policy (or which, at the other extreme, seeks to have everybody in the hands of public clinics). Now, the advantages of the HCDS will not be fully apparent until work is completed in this and the final chapter. But the point, with respect to qualification mechanics, is this: the HCDS, as constituted, can be used to supplant not only the unwieldy and incoherent federal schemes like Medicare and Medicaid (and the raft of categorical programs now supported by the federal government), but can also complement the private coverages secured for employees by public and private employers. That is, the HCDS, given a mandate for *total* community health care, can promise positive impact on the entire structure of medical practice and costs in any community (and, by extension, in a region or the nation). The reason for this is that the population as a whole—the employable population as well as the set of indigents —can benefit from the type of intelligent analysis the HCDS performs as a matter of course, particularly with regard to the definition of different coverage modalities and the rationalized referral system (both of which are discussed in subsequent sections). In short, then, not only should an HCDS keep its fundamental qualification threshold flexible, but so should its definition

*In short, the number of individuals who would become eligible—or be rejected—in the face of incremental increases or decreases in qualification criteria is the target of the demand functions.

of its client population be allowed to extend all the way from a restricted population of socioeconomic indigents to an unbounded population of residents in a community or region as a whole.

The assertion being made here, then, is this: a *national* health care delivery system should be comprised of a collection of local HCDS units of the type outlined in these pages. This would mean that only where a client's socioeconomic and conditional (medical) attributes explicitly recommend it would he be placed into some sort of insurance (e.g., Blue Cross) scheme. For other clients, irrespective of whether they are public charges or employed by some governmental or commercial enterprise, some sort of directed and monitored delivery modality is preferable, perhaps the block or HMO modalities discussed in the next section. The HCDS would not, in this instance, socialize medicine. Rather, as will be seen, every effort is made to maximize the contribution of mainstream providers. Nor would the HCDS system put the private insurance systems out of business. Rather, the use of some discretion in posing clients for enrollment in prepaid plans would be of considerable appeal to insurers who are now being asked to take clients despite actuarial or conditional equivocations. As arguments are completed in these pages, it is hoped the reader will be able to see that the extension of insurance coverage on a blanket basis could only accelerate the increase in costs of medical services (and increase the instances of abuse that characterize the Medicaid and Medicare systems). By the same token, an effort to socialize medicine—to bureaucratize the entire health care process—is perhaps even more unattractive from either the indigents' or the taxpayer's perspective. Therefore, as for the matter of qualification, further analysis and study of the HCDS potential might very well lead to the situation where local programs offer their service to local employers (public and private) and extend the relevant client population far beyond the bounds of the indigent. (A defense of this assertion appears later on).

With these broad strategic qualifications in mind, we may rather quickly move through the remaining logic of Chart 4.1. The key component is box 2.1, where it is asked whether or not the prospective client has been accepted into the program, whether he indeed meets the current eligibility (qualification) criteria. Let's first of all see what happens when he is rejected, by following the no path from box 2.1. Initially, attention is paid to the transfer potential—has the client a condition for which some other social service agency is responsible? If the answer is yes, the client is referred to the appropriate agency (which means that the client would enter some other social service program through the transfer access modality with the HCDS initiating rather than receiving the transfer). Also to be asked is whether the client's condition is *reportable.* A reportable condition is one which demands the attention of some other authority (e.g., a contagious disease of a certain sort; drug addiction). An alert would be initiated to the appropriate authority and followed up to see that the client did indeed make contact. This is the purpose of the logic beginning with box 2.1.7.

Let's now go back to box 2.1.3 and see what happens when there is no other agency which should be alerted. If we follow the no path we arrive at box 2.1.4 which asks if there is a medical condition for which a referral might be made. In this case, the client is in need of some treatment, but did not meet the socio-economic qualifications of the program. But, where resources permit, there is the option to exercise a *courtesy referral* on his behalf. A courtesy referral, in its broadest implications, means that *any* community resident may ask to be referred to a medical provider for any condition requiring treatment. There are two reasons why the courtesy-referral mechanism can be important. First, even well-schooled citizens may sometimes have difficulty locating an appropriate provider. This is especially true of young people who have had few medical problems or of new residents in the community who have not yet established a relationship with a family physician. In other cases, it may be that the medical condition is an uncommon one, and the client may simply have difficulty in locating an appropriate specialist. Therefore, the courtesy-referral mechanism is again an instance of a public agency trying to be of service to the community as a whole rather than simply to social casualties. Second, as will be seen in the last section, referrals made by the HCDS system are *rationalized* and consider geographic, quality, cost and distributive criteria. Therefore, when referrals are made through the HCDS-referral mechanism, both the client's interests and those of the community as a whole are equally well served. HCDS-initiated referrals tend to take business away from unfavorable providers in a community and support the more cost-effective. Thus, when citizens at large ask the HCDS for a referral, it is essentially an action that can help improve the community's general medical facilities, which eventually rebounds to everyone's benefit.

Returning to the central question box (2.1), let's see what happens once a client has been accepted into the system, by following the yes path. The first box (2.3) tries to identify any third-party payors which might have responsibility for certain medical conditions the client might have. For example, certain military-related conditions may be the proper responsibility of the Veteran's Administration, and, therefore, the HCDS should not be responsible for their treatment. In other cases, there may be certain categorical programs which take care of special conditions and thus absolve the HCDS of financial (but not necessarily referral) responsibility. Thus, it is important to identify any conditions which are covered by third-party payors, this is in the obvious interests of conserving HCDS resources for those conditions for which there is no alternative coverage available. Thus (box 2.3.1), any such coverage restrictions are flagged on the client's file, and the client is thus made aware of those conditions which HCDS will not support. In short, the HCDS, with every new client, seeks always to determine a *residual responsibility*. Finally, box 2.3.2 asks about any *copayor* provisions that might apply to the client. Obviously, a copayor provision is attractive in that it would ask the client, to the extent possible (given the socioeconomic qualification criteria), to contribute something toward the treatments he requests or receives. This copayor provision argues against indiscriminate or excessive use of the HCDS coverages and also discourages the formation of Medicaid mills or other exploitative provider enterprises. The exact copayor provision, for any client, would be related actuarially to his economic status and his projected use of the system (given the medical screening data generated in a subsequent section of the HCDS system), and would top out at the point where the individual becomes effectively indigent. That is, copayor logic implies a limited deductible contribution from the client.

The last of the access-qualification components is the provision for an *appeal* for rejected clients. This implies that the HCDS system will support some sort of *internal ombudsman mechanism* which will review rejections (on request) for any irregularities or abrogations of due process. The point of the ombudsman process is to minimize civil legislation by giving a client objective resolve of his rejection and, derivatively, to provide an objective evidentiary base should a client elect to pursue his cause through the courts. A confoundation can occur when a client received treatments on a contingency enrollment basis and was subsequently rejected. In this case, the client becomes answerable for the resources expended on his behalf and becomes intelligible to the system as an account receivable. The same thing would be true of copayors who fail to meet their obligations. But when a client has been finally rejected, after any appeal that might have been evoked, he is assigned a Code-4 (box 2.2.4) and released from further processing.* This last provision concludes the operative logic of the access-qualification mechanism of the HCDS.

THE COVERAGE ASSIGNMENT PROCESS

Once a client has been accepted into the HCDS system, two ancillary administrative functions are exercised. The first is the *screening* process; the second is the *assignment* of the client to a particular coverage (delivery) modality. The screening function simply seeks to develop a data base about the medical condition of the client, much as the qualification process developed a data base as to his socioeconomic (and behavioral) attributes. Thus, the screening function serves as an input into the coverage assignment function, as Chart 4.2 indicates.

Chart 4.2 begins at the point where the client is released from the access-qualification subsystem (from box 2.3.2 in Chart 4.1). The first question to be asked is whether an adequate medical profile is available. A medical profile would enable us to suggest something about the nature and schedule of treatments the client might be expected to require and, hence, something about the demands that client might be expected to make on the HCDS. No screening, per se, would be required of transient clients; rather, any acute condition requiring treatment would be noted and the client referred to an appropriate provider. But a medical screening might be indicated for clients who are expected to remain with the system. In general, two conditions tend to recommend a client to the screening function: (a) where

*The use of client codes will be discussed in the next chapter.

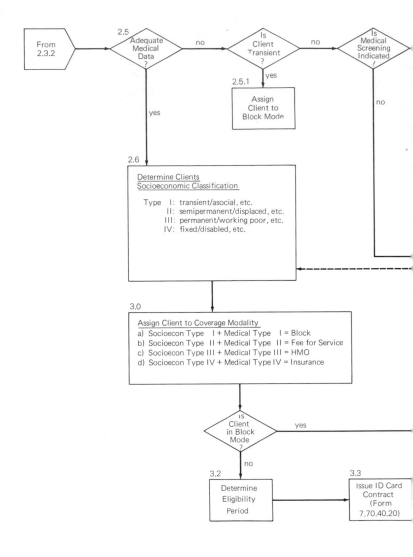

CHART 4.2 / Screening and coverage assignment subsystem

resources permit, a complete medical screening might serve the purposes of *preventative* medicine, seeking to identify incipient conditions and thus suggest preemptive treatments; (b) wherever a client's medical profile is incomplete, or where there are ambiguities about the medical category in which he should be

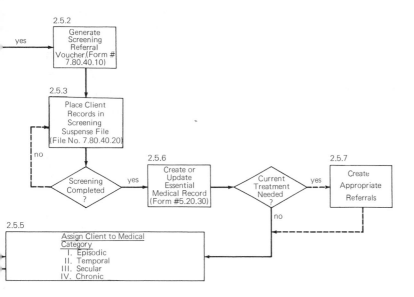

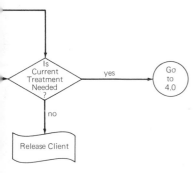

placed (box 2.5.5), a screening must be performed. The latter condition serves a distinctly *administrative* purpose.

The relative simplicity of the components of Chart 4.2 covers a multitude of complexities, for the coverage assignment decision is among the most crucial to be made anywhere in the health care delivery process.* Looking at the chart, we can see that the coverage assignment decision is made in box 3.0, given two prior assignments made in box 2.6 and box 2.5.5. Again, the assignment of a client to a particular medical (conditional) category is made on the basis of his medical records and/or from information obtained from a special medical screening. The assignment of the client to a particular socioeconomic category is made on the basis of the information obtained during the qualification review process (box 2.0 in Chart 4.1). The assignment of a client to a particular coverage (delivery) modality depends jointly on his personal and conditional attributes. As an ancillary point, the different coverage modalities determined in box 3.0 also determine, in large measure, the different eligibility intervals clients will be assigned in box 3.2.

As has been explained earlier, the concept of developing different coverage modalities to respond to different client and conditional attributes responds to the central dictate under which any social service program operates: work toward a favorable tradeoff between system effectiveness, efficiency and client dignity. Axiomatically, not all clients—and not all medical conditions—can be equally well treated by any given coverage (or delivery) modality. Thus, there should be as many unique coverage modalities as there are unique client or conditional classes.

A defense of this assertion—and showing how the HCDS actually developed different coverage modalities, client categories and conditional classes—appears in the pages that follow.

Defining Coverage Modalities

Again, the purpose of this aspect of the system is to ensure that the system manager is able to make best use of resources by assigning clients to that particular coverage modality which will allow him to receive required health services at the lowest expected cost to HCDS. Though the logic is fairly simple, the decisions made in this regard will have considerable impact on the overall cost-effectiveness index a health care program will be able to achieve.

The assignment logic involves, first, a specification of the coverage modalities available to the system; the unique attributes of each modality must also be set out. Second, the socioeco-

*And in other social service contexts as well.

nomic and behavioral criteria must be established which will serve to determine into which category a client should be placed. Third, there must be a specification of the medical (conditional) criteria which should determine assignments. Finally, the medical and socioeconomic criteria must be combined in a way which will make clear what are the *congruent* assignments, given the modality attributes and the socioeconomic/medical criteria set out.

There are basically three coverage modalities available to the HCDS:

1. *Prepaid:* Under this modality, a one-time or periodic payment is made to an agency which will then guarantee a client access to unlimited contacts for treatments which are covered under the contract terms for the duration specified. There are basically two types of prepaid submodalities of concern here:
   a. The Insurance modality indemnifies the client against all (or some portion) of health care costs with a limited coverage schedule in terms of both treatments and aggregate payments.
   b. The HMO modality makes available a specific provider location(s), which will, for a prespecified fee, provide all agreed services for that client.

2. *Limited-Term Fee-for-Service:* Under this modality a client would be indemnified by HCDS for some specific period, against a limited or unlimited schedule of treatments. The client might, under some circumstances, be a copayor (such that indemnification is only partial).

3. *Block:* Under the block concept, a physician, group of physicians, a hospital or a clinic, etc., are paid a block fee which serves as prepayment for particular treatment for some number of unspecified clients.* The amount of the prepaid fee would be determined as some portion of the standard rate (with the precise proportion being determined by factors such as the amount of time and numbers for which prepayment is made). As a rule, the block modality should be used to cover only well-specified conditions which are expected to appear among clients with some frequency and

*Note that the *centrix* concept, discussed earlier, would suggest the approximately optimal locations for block providers.

for which purchase of mainstream care or other modalities would be infeasible or more expensive than necessary: medical screening, detoxification, delousing, one-shot (episodic) emergency or acute treatment for ambulatory clients, etc.

Table 4.1 sets out key attributes associated with each of the coverage modalities. Examining the implications of the table in a bit more detail will help indicate why the definition of different delivery modalities is so important. *Accessibility* refers to the geographic distribution of acceptable providers under each of the modalities. The HMO and block modality each have restricted service locations; under the insurance and fee-for-service modalities, the client is not restricted as to provider locations.

With *continuity* there is concern about the modalities' ability to provide a continuous and integrated stream of services to the client. Thus, the HMO scores high on this account, while, because the block modality is a one-shot affair, the criterion is not applicable. Continuity for the insurance and fee-for-service modalities is really indeterminate; it would be high if the client used a family physician, low if he did not maintain a consistent medical provider. At any rate, under these latter modalities, the client actually determines the extent of continuity.

Overhead impact refers to what can be expected about administrative overhead on the HCDS from each of the modalities. Were the insurance modality used, fiscal demands arise only as a client's term expires and are usually inaugurated by an invoice from the insurer. Similarly, the insurer will have records which, in effect, "track" the patient's movements in the system. To the extent that such records would be available, the patient-tracking function is performed externally; if these records were not available from the insurer, then the patient-tracking problem would be

TABLE 4.1 / Coverage Attributes

Attributes	Insurance	HMO	F-F-S	Block
1. Accessibility of Providers	high	low	high	limited
2. Continuity of Service	indeterminate	high	indeterminate	none
3. Overhead Impact on MMIS	low	low	high	moderate
4. Decision Implication for Client	high	moderate	moderate	low
5. Cost Function	fixed	fixed	variable	complex
6. Provider Selectivity	high	high	moderate	low

most difficult, for the normal invoice-based tracking method would not be appropriate. Therefore, it is urged that whenever contracts are negotiated with insurers, the HCDS be assured of access to the client's history during the period. If such assurance is given, then the administrative overhead implications of this modality are low, as was indicated in the table. A similar situation holds with respect to the HMO. Fiscal demands on the HCDS arise only periodically and in response to an invoice from the HMO itself. The HMO would also maintain internal records which suffice as patient-tracking data. When these are available, the system would simply have to "borrow" them to perform this function. However, should such data not be made available, then the construction of a patient-tracking profile and the maintenance of a medical record would be impossible, for all events take place internal to the HMO itself. Therefore, as with the previous modality, it is critical that before contracts are let access is assured to the internal data (and that the system be structured to incorporate the insurer's and HMO's internal data). The fee-for-service modality, on the other hand, makes considerable fiscal and tracking demands. There is a short billing-invoice cycle, and the system has sole responsibility for the client-tracking data. It will be noted that the data processing logic reflects this worst of all possible worlds, and the basic record keeping and referral and reimbursement provisions are capable of handling the demands of this modality.* In fact, the system considers this the "normal" situation. Finally, the block modality is expected to make moderate demands on the HCDS. Unlike the previous modality, fiscal demands arise only periodically and may be controlled by the number of contacts which are prepaid (and the length of the contract period). However, the HCDS will still have the sole responsibility for keeping the client profile, though in some cases the profiles may be updated periodically rather than continuously.

Decision implications for the client refers to the extent to which the client himself is responsible for his own health care decisions. The insurance modality demands the most of the client, for the system manager or the HCDS staff will have only slight opportunity to audit his real-time progress (as opposed to the fee-for-service modality which allows auditing of the patient contact-by-contact with only a slight lag). The HMO modality

*See the last section of Chapter 5.

because of the high continuity and single-source characteristic would place only moderate demands on the client. Obviously, demands are lowest under the block modality, where decisions as to health care are all made *for* the client. Thus, to wisely use the insurance modality would demand some sophistication from the client, while the block modality would demand least, with the fee-for-service and HMO modalities making moderate demands.

Cost characteristics are of obvious importance. The insurance and the HMO modalities both carry a fixed cost, that is, independent of the rate of utilization of services by the client. For the fee-for-service modality, aggregate cost of serving a client over some period is entirely variable with his rate of utilization of providers. The block modality presents a slightly more complex picture. The cost of serving any individual client is a product of the number of contacts he has times the average cost per contact (i.e., the capitation rate at which the block provider was funded). Thus, for any single client, costs are fixed within the limit set by the number of contacts prepaid by HCDS. However, this number itself varies with the rate of utilization of the provider's services (any single client's costs will vary with his utilization).

Finally, there is the matter of *selectivity.* This attribute refers to the modality's ability and motivation to restrict access to clients on the basis of medical, socioeconomic or administrative criteria, all with respect to individual enrollees. In short, the concern here is with the severity of the criteria these various modalities might employ. While precise criteria are not available, it should be clear that the insurance modality would stand to gain the most by a strict set of criteria (as their profit depends directly on the positive difference between the premium paid and the rate of utilization of covered services by a client). Thus, they must be statistically (i.e., actuarially) satisfied that a prospective client won't seriously erode their fiscal situation, beyond the calculated risk of any client's suffering an accident, etc. It can be expected, then, that the insurers available under this modality would apply some sort of medical criteria with respect to individual enrollees (or perhaps set restrictive limits on any group contract). The HMO will be slightly less selective (expectedly), for unlike the insurance company (for which every client represents a variable cost entity), the HMO has a significant fixed-cost component which must be amortized over a large number of clients, such that even an excessive user of facilities still contributes a positive amount to overhead. However, the HMO's selectivity must be expected to vary positively with the rate of utilization of their facilities. Underutilized HMOs may be less selective than those having an excess demand given their infrastructure. Again, then, medical criteria are expected to play some part in an HMO's acceptance of a prospective client.

The fee-for-service provider is likely to be less concerned with medical criteria than with socioeconomic or administrative implications. Obviously, his income will vary directly and positively with rate of utilization (he might actually prefer clients with some significantly high probability of being ill, such as elderly people). On the other hand, the private practitioner may have very definite ideas about the socioeconomic *class* he wishes to serve. Thus, he may restrict services to those clients who are demonstrated as being acceptable to him. Finally, his assuming of a client might also depend on the administrative overhead he will have to absorb. To the extent that the administrative demands of HCDS on the individual practitioner can be minimized, his acceptance of clients will be increased (subject, of course, to the socioeconomic criteria).

As for the block provider, certain factors act to make him expectedly least restrictive. For one thing, we can expect to develop block service contracts with a fairly large number of special clinics (e.g., the free clinics in inner-city areas, or those essentially charitable clinics established by private social agencies and religious groups, etc.). These clinics, along with the ethnic clinics, actually exhibit a preference on many occasions for clients with a discouraging socioeconomic profile. Also, we could expect to develop block contracts with certain government-sponsored categorical clinics for whom discrimination is not possible. Finally, the possibility of gaining guaranteed fiscal solvency might appeal to clinics established by younger, less affluent physicians; in some cases, block contracts may actually be a motivation to the establishment of such clinics in areas where adequate services do not already exist. At any rate, the block modality will provide access to those clients a priori least acceptable to the other modalities.

It is clear that, from a strictly economic standpoint, assignment of a client to a given modality would be made *rationally* by considering the function presented in Figure 4.3. Quite simply, as already suggested, it is expected that the *economy* of the modalities varies directly with the frequency with which events demanding medical service occur and with the expected costs of these events. To the extent that any prospective client is expected to have a series of frequent, expensive conditions, the prepaid modality becomes most appropriate; if a client is expected to have only widely spaced single episodes of short duration and requiring fairly common (as opposed to specialized) treatment, then the block modality presents itself as the best assignment.

This type of decision mechanism is not likely to be initially of much use to the system manager, for the exact nature of the function will not be a priori calculable, nor will he (at first) be able to predict with too much accuracy what kind of medical future clients are likely to have. For these and other reasons, he must make coverage assignments using *qualitative* (or logical) analysis, trying, however, to gradually build up an empirical data base which will allow him to make more precise, better disciplined assignments at some future point in time (using a quantitative model like the above). Fortunately, there are some very effective logical instruments which can help the system manager make expectedly most efficient coverage assignments. These will occupy the pages that follow and deal, essentially, with the explication of certain socioeconomic and medical classifications which, when set up in association with the properties of the coverage modalities just discussed, should enable the making of *congruent* assignments even in the absence of quantitative data (a congruent coverage assignment being that which promises to make best use of the HCDS resources provided by various funding sources). Again, however, it ought to be stressed that a careful audit be kept of decisions taken on the basis of the logical instruments to be introduced, for such audits will eventually allow displacement of the logical instruments with more precise quantitative ones (or

FIGURE 4.3 / Cost characteristics of modalities

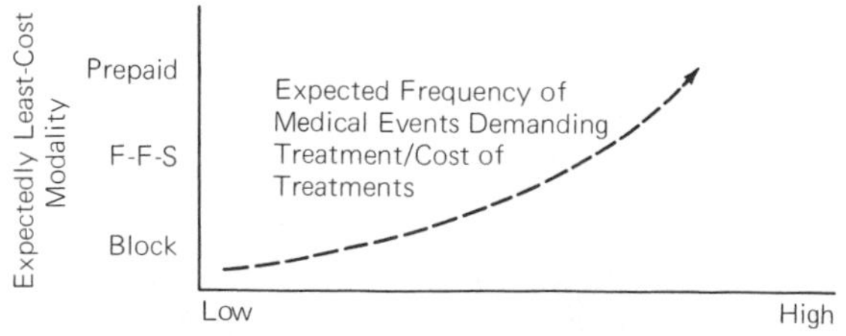

might even validate the logic being used so that a shift in decision methods need not be made).

Socioeconomic and Conditional Classifications

Because the major concern is with assigning a client to the *right* modality for him (and for the economic integrity of the HCDS itself), any socioeconomic criteria or classifications developed should have direct reference to the attributes of the various coverage modalities earlier defined. The criteria which might be used are those listed in Table 4.2. Each of the "values" assigned the various social classifications have definite administrative implications for the HCDS system, as follows:

> *Residency Status:* The values assigned this dimension indicate the expected permanency of residence within the geographic area of eligibility. Transients are those who have no history of residence and a low probability of remaining beyond a short period of time (e.g., a period longer than six months). The semipermanent resident would be one whose expected period of residence would be between six months and a year (e.g., seasonal workers, students, temporary employees of local firms); or his period of residence may be indeterminate, depending on the emergence of an opportunity to move somewhere else . . . he is highly mobile. The long-term resident would be one with a low probability of moving, perhaps indicated by a significant period of residence and a more or less permanent family or employment

TABLE 4.2 / Socioeconomic Classifications

Criteria	I	II	III	IV
Residency Status	Transient	Semi-permanent	Long-term (secular)	Fixed
Economic Profile	Unemployable	Structurally unemployed	Working poor	Retired/ disabled
Social Profile	Regressive	Displaced	Deprived	Impaired
Cognitive Class	Pathological	Competent	Assistable	Senile/ retarded
Mobility	Contained	Moderate	High	Immobile/ nonambulatory

situation. The fixed resident is one for whom the possibility of moving simply does not exist (e.g., the elderly, the immobile).

Economic Profile: The unemployable would be an individual whose characteristics (endogenous, intrinsic) make his securing of employment very improbable. The structurally unemployed individual is one who is temporarily out of work because, usually, of economic or technical factors beyond his control. The working poor are those who are employed at an extremely low level of subsistence. The retired or disabled individual is secularly unemployable because of chronic limitations.

Social Profile: The regressive individual is one whose basic behavior is asocial, self-destructive or otherwise inimical to accepted standards (e.g., the alcoholic, the addict, the hobo). The displaced individual is one suffering a temporary and remedial condition (e.g., the newly widowed husband with children; those injured in industrial accidents; those undergoing job retraining). The deprived individual has an educational or social background (e.g., endogenous factors) that makes him unable to compete successfully in the economic arena (e.g., the casual or unskilled laborer) or hampers him socially (e.g., the immigrant). The impaired client suffers a permanent or chronic disabling condition.

Cognitive Class: This refers to the individual's intellectual capacity. The pathological is one who is potentially self-destructive or asocial (e.g., the "strung out" addict). In contrast, the senile/retarded individual is intellectually incapacitated, but not dangerous either to himself or others (e.g., he is tractable). The competent case indicates an expected ability to make proper decisions regarding necessary care and when and where it should be provided. The assistable individual, on the other hand, is one whose social or educational deprivations indicate a need for continual or periodic guidance, advice and auditing of his medical experiences and needs.

Mobility: This refers to the individual's ability to transport himself from residence to medical provider. The contained individual would have to have an escort (perhaps a legal authority, etc.) and his movements may be constrained. The individual with moderate mobility might be able to make use of low-cost public transportation, but has no automobile or insufficient funds for a taxi, etc. The immobile individual would require both an escort and possibly special transportation (e.g., an ambulance). The high mobility individual has no limitations.

As with the socioeconomic classifications just set out, medical classifications need relate only to the question of assigning individuals to coverage modalities. Therefore, the four classifications in Table 4.3 are sufficient. It should be noted that medical "events" of interest here are only those for which HCDS has a coverage responsibility (e.g., those for which no third-party payor is responsible or for which the client is not otherwise indemnified). Given the "values" for the criteria determining medical classifications, then, the following are suggested:

The *episodic* individual will be one suffering an acute condition of short duration which is unlikely to recur again in the forseeable future. Aside from this single condition, the individual is expected to remain in excellent health, or at least not require any predicted treatment with any calculable frequency (treatment, that is, for which HCDS is deemed responsible).

TABLE 4.3 / Medical Classifications

Criteria	Episodic	Temporal	Secular	Chronic
Nature of Medical Event(s)	Single Condition	Set of related events or replication of a single condition	Wide range of potentially diverse conditions	Focussed array of conditions
Expected Duration	Acute	Short-term (e.g., six months)	Indeterminate (variable in duration)	Chronic (or terminal)
Expected Frequency of Appearance	Random (one-shot)	Clustered in a specific interval of time	Stratified through time	Constant

The individual falling into the *temporal* classification would be one suffering from a set of highly correlated (related) symptoms—or from a recurrent condition—of predictably short duration (with the necessary treatments expected to fall into a well-defined interval). On a broader scale, the temporal would be that assigned to the currently healthy individual who applies for general HCDS coverage in the absence of any conditions requiring immediate treatment. In short, the temporal classification should be considered the enrollment "null" hypothesis.

The individual falling into the *secular* category would be one for which HCDS has broad and long-term coverage responsibility, and who is deemed likely to have a wide range of diverse conditions requiring treatment, this likelihood being derived from contextual or actuarial data (e.g., the child of a working poor family with a history of illnesses; a person approaching middle age with a heart condition or obesity). In short, the secular case differs from the temporal in that there is deemed to be a significant probability of rather frequent use of medical services for the former over a relatively long span of time.

The *chronic* category would be represented by those with long-term (possibly terminal) conditions of paramount importance (e.g., the cancer victim; the geriatric case; the kidney patient; sufferers of Bright's disease, etc.).

It is important that the HCDS processes identify in the initial stages of client enrollment the data which would permit an assignment of an individual to one or another of these medical classifications. Where an assignment is not immediately evident or producible from the client's medical history, the medical screening provision is at our disposal.

Congruent Assignments

The socioeconomic and medical classification models developed in the preceding sections—along with the definition of the various coverage modalities and their attributes—give the basis for suggesting the nature of the coverage decisions which *should* be made by the HCDS system manager under the criterion of congruence (demanding the assignment of any prospective client to that particular coverage modality which promises to be effectively optimal, given his socioeconomic and medical classifications). There are several different conditions under which decisions become obvious. We shall deal with them in order of complexity.

First, whenever a client appears who substantially meets the criteria for one or another of the basic socioeconomic classifications in Table 4.2, the system manager should make assignments according to Table 4.4. As the table shows, there is a *vector of congruence* which easily develops when socioeconomic classifications are matched with the attributes of the various coverage modalities. The two areas outside this vector of congruence would represent either infeasible or uneconomical assignments.

Uneconomical assignments would occur when an individual is assigned to a coverage modality that would carry a positive opportunity cost (that is, one which is overly expensive). Thus, it would be uneconomical to assign a type III socioeconomic client to the block modality, for from the permanency of his residence and the fact that he is a working poor, etc., it is possible to assume that he will have rather significant and long-term need for medical services (hence, he should be put on some sort of fixed-cost coverage, either the insurance or HMO modality). But his other properties enable the identification of the HMO as the *most* probably optimal assignment: (a) he is highly mobile, so he is assumed capable of making his own way to the limited number of service centers which an HMO makes available (unlike the effectively immobile type IV individual); (b) the fact that his background marks him as possibly unable to make optimal medical decisions for himself, the partial control of the HMO makes this attractive with respect to the insurance modality (where the individual himself must make all decisions).

Second, like the socioeconomic classes, medical classifications have direct implication for the assignment of coverage

TABLE 4.4 / Socioeconomic/Coverage Modality Correlations

Socioeconomic Classification	Congruent Coverage Modality			
	Block	Fee-for-Service	HMO	Insurance
I	X			
II		X	(Infeasible Assignments)	
III	(Uneconomical	Assignments)	X	
IV				X

TABLE 4.5 / Medical Classification/Coverage Modality Correlations

Medical Classifications	Congruent Coverage Modality			
	Block	Fee-for-Service	HMO	Insurance
Episodic	X			
Temporal		X	(uneconomical)	
Secular	(uneconomical)		X	
Chronic				X

modalities, perhaps even more direct. The correlations we want to suggest are those of Table 4.5. The logic behind these assignments is fairly evident. Initially, the episodic-block correlation occurs because this is the most economical way to have a one-shot condition treated; sending the episodic to a block provider is simply making use of already "sunk" (fixed) payment. A temporal client becomes an excellent candidate for fee-for-service coverage for his condition is a replicative one (or more complex than those possibly treated by a block clinic) and is of fixed duration. Therefore, provision of a prepaid coverage modality would be uneconomical, and making him go to a block clinic would put an undue stress on the client himself (e.g., he can effectively use a local physician). The HMO-secular correlation takes advantage of the broad range of in-house services provided by HMOs and answers the fact that the secular patient will not usually have need for the *specialized* services that the chronic client will demand (services which may be provided under the insurance modality but might be at least partially unavailable at the traditional HMO). Thus, had the HCDS manager *only* medical data available, these are the suggested assignments.

Now, from what has been done in the immediately preceding sections, it is clear that whenever a client falls substantially into either a specific socioeconomic or medical classification, the assignment of that client to the expectedly optimal coverage modality is a clear-cut decision. Hence Table 4.6. It is impossible to tell, a priori, just how many clients will appear who fall neatly into one of the established categories and hence permit the use of this particular assignment (decision) table. It is hoped the number of clear-cut cases will be rather large.

We must now consider the more complex cases where a client is associated clearly with a particular medical classification, but perhaps exhibits properties of a mixed socioeconomic nature (that is, his properties are drawn from one or more socioeco-

TABLE 4.6 / Table of Congruences

SocioEconomic Classifications	Medical Classifications			
	Episodic	Temporal	Secular	Chronic
I	BLOCK			
II		F-F-S	(uneconomical)	
III	(uneconomical)		HMO	
IV				INS.

nomic classifications rather than being neatly associated with only one). When a client exhibits a complex set of socioeconomic properties in conjunction with a given medical classification, the assignment of that individual to a specific coverage modality becomes a *complex* decision. A decision has to be made, however perhaps with reference to Table 4.7. It can be seen that this table would dictate that essentially the same assignments be made as those prescribed by Table 4.6, but it points out some few areas where socioeconomic properties would take precedence over the purely economic aspects of the various medical classifications.

For example, the pathological cognitive condition demands assigning the individual to a block provider irrespective of his medical classification, for his behavioral characteristics will probably exclude him from a private practitioner and perhaps even the HMO. Similarly, a long-term resident (e.g., type III or IV) under the episodic medical classification might as well be assigned to the fee-for-service modality and allowed to visit something other than a block clinic. Because he is a long-term resident, he is likely to have need of services again at some point in the future; by assigning him a fee-for-service classification, he need not be completely rescreened or robbed of his ability to go to the provider of his choice (as done, for example, with all pathological cases irrespective of their residence status). Much the same logic would hold in allowing the working poor (i.e., type III under economic profile) episodic to have a fee-for-service classification, for his stable employment would indicate a certain reliability on his part and some probability of future use of the HCDS services. Hence, eliminated are the necessity for rescreening and other administrative overhead operations, and the individual is possibly saved any embarrassment that a block provider situation might cause. Note, however, that were this working poor individual also equipped with a type III social profile—deprived—a

TABLE 4.7 / Decision Table: Complex Assignment Cases

SocioEconomic Properties	Episodic	Medical Classifications		Chronic
		Temporal	Secular	
1. Residence Status				
I	BLOCK	BLOCK	F-F-S	F-F-S
II	BLOCK	F-F-S	F-F-S	F-F-S
III	F-F-S	F-F-S	HMO	INS
IV	F-F-S	F-F-S	INS	INS
2. Economic Profile				
I	BLOCK	F-F-S	HMO	HMO
II	BLOCK	F-F-S	F-F-S	F-F-S
III	F-F-S	F-F-S	HMO	HMO
IV	F-F-S	F-F-S	INS	INS
3. Social Profile				
I	BLOCK	BLOCK	HMO	HMO
II	F-F-S	F-F-S	HMO	INS
III	BLOCK	BLOCK	HMO	HMO
IV	F-F-S	F-F-S	INS	INS
4. Cognitive Class				
I	BLOCK	BLOCK	BLOCK	BLOCK
II	F-F-S	F-F-S	F-F-S	INS
III	F-F-S	F-F-S	HMO	HMO
IV	F-F-S	F-F-S	F-F-S	F-F-S
5. Mobility				
I	BLOCK	F-F-S	INS	INS
II	F-F-S	F-F-S	INS	INS
III	BLOCK	F-F-S	HMO	HMO
IV	F-F-S	F-F-S	HMO	INS

block assignment would be recommended due to the possibility that a private physician might refuse him service because of his low status or because of his inability to make a proper decision about what provider to see, etc.

Every neat decision process, however, has its confoundations. The assignment logic set out in Table 4.6 (and the more complex assignment criteria of the last table) result in the expectedly best use of HCDS resources and also incorporate the criteria of effectiveness and client dignity. The very individuals we might most wish to place into the insurance or HMO coverage modalities (e.g., the medically chronic or the type IV socioeconomic individual), might be precisely those whom the insurer and the HMO might want to exclude. However, the treatment of such individuals under a fee-for-service modality could be very expensive or inappropriate. Therefore, it is recommended that every effort be made to realize congruent assignments in practice, even if this means paying a negotiated rather than standard rate to the insurer or HMO. The limit on such rates should be calculated with respect to some expected value for the cost of maintaining the individual in a fee-for-service modality, and the rate negotiated by the insurance coverer or HMO (probably derived from actuarial studies) should, of course, be more favorable than the cost projection the HCDS calculates.

An alternative is to anticipate such situations by performing actuarial studies in-house and developing a sliding rate schedule for prepayments based on expected frequency of medical events and expected costs. At any rate, given these few qualifications, the coverage assignment process seems to have a critical place in the usual social service delivery system, and the logic outlined here will help others incorporate this feature in emergent programs or perhaps as an addendum to existing programs.

Eligibility Intervals

Once clients have been assigned to a particular coverage modality, the client classifications (Table 4.1) may be used to help make a determination about the length of the eligibility period the client should be assigned. This is the last of the tasks to be performed in the screening and assignment subsystems. Eligibility intervals are an important determinant of overall system effectiveness and efficiency. When intervals are short, such that clients are required to be requalified frequently, a decrease in the incidents of client fraud or abuse occurs (e.g., using system coverage when the client is no longer qualified), but only at the expense of increased administrative overhead, for client requalification is expensive in terms of clerical and personnel costs. On the other hand, when seeking to minimize administrative overhead by making only infrequent requalifications (allowing clients to have an extended eligibility interval), the expected cost of abuse increases. Thus, the decision as to the eligibility schedule to impose on the client population becomes another of the many tradeoff issues the system manager must be prepared to resolve.

Client classifications of the type developed in Table 4.2 in large part answer for the terms of the tradeoff. Different client classes naturally point to different eligibility intervals, perhaps according to a scheme like Table 4.8. The vector of congruence here suggests generally that eligibility periods should increase as the client moves through the various classifications. Those clients who would fall into category 1 of Table 4.2 thus should be requalified each time they seek to use the system's services, whereas clients falling into Category IV are expected to remain eligible over the extremely long run (being, as they are, chronically indigent). Off the major diagonal (the vector of congruence) are the expectedly unsatisfactory eligibility assignments. Assignments to the right of the vector would be cases where the risk of abuse is too significant given client characteristics. On the other hand, the assignments made to the left of the major diagonal indicate situations where we accept excessive overhead costs relative to the client's probability of abusing privileges. For example:

> Assigning a type I individual a significant interval of coverage might lead to the situation where he would seek to "sell" such privileges.

> Assigning a type II individual a long-term coverage might discourage that individual from making a geographic move to another area where employment is available.

> Assigning a type III individual to an indefinite coverage might cause disregard of critical changes in the structure of clients' social or economic properties.

Obviously, the specific durations listed in this table are just arbitrary; any system manager would manipulate them as circumstances dictate.

TABLE 4.8 / Eligibility Intervals Relative to Client Classifications

Client Classification	Eligibility Interval			
	0	6 mos.	18 mos.	Indefinite
I	X			
II		X	(Risk of Abuse	too Significant)
III	(Overhead too High)	Expectedly	X	
IV				X

We must, however, raise a qualification here. Recall, particularly, the earlier argument that every system manager should seek to exercise control over the qualification threshold, so that client population may be quickly adjusted to changes in the system's per capita resource level. The rationale was that it is extremely difficult to establish a proper resource expenditure rate, such that service levels remain constant throughout a fiscal period (and the system neither underspends nor overspends relative to the client population). This difficulty is obviously greater at the formative stages of a new social service program or in the face of significant local politicoeconomic perturbations.

For obvious reasons, it is impolitic to be constantly changing eligibility requirements when a large number of clients have already been admitted. There could be expected to be a vocal reaction were the system to suddenly withdraw coverage from previously accepted clients. Therefore, until a social service program has matured, it is important that eligibility periods for *all* clients be kept short enough so that admission could be adjusted with frequency. In short, if the system manager is to be able to obtain a normative expenditure profile*—and if he is to be able to avoid adverse reaction from clients who having been given long-term eligibility intervals are now dismissed from the system —then long-term obligations must be kept to a minimum during the initial phases of implementation (when the system is not equipped with the data necessary to make effectively accurate estimates of the demand for medical services under the broker system).

In short, then, the eligibility recommendations made under the previous table should *not* apply to the first phases of system implementation. Rather, it is suggested that type II, III and IV clients all be awarded quarterly (three month) periods of coverage so that, if a condition of overdemand and overexpenditure results and certain clients have to be dismissed, the adverse reaction will be minimized. Thus, while the system manager is still in a learning state, he will leave himself the greatest amount of latitude in adjusting eligibility thresholds on the basis of actual expenditure-demand conditions as they emerge and change. In the initial stages of operation, then, he is forced to sacrifice some economies which would result from longer intervals for certain individuals in order to minimize the risk of seriously embarrass-

*See the first section of Chapter 5.

ing the program should demand projections be exceeded or should resource expenditures threaten to run the system out of funds before the end of the fiscal year. This same logic would hold true for either general or categorical funds, or for mature systems facing significant changes in their operating environment.

Keeping initial eligibility intervals short also allows the system manager to husband resources so that those cases most seriously demanding treatment are allocated resources first (under either general or categorical budgeting). If, subsequently, a condition of underexpenditure emerges, then the next quarter could have a larger number of openings by simply accepting clients from a lower priority level. Thus, the system manager will have to equip himself with a priority scheme to determine a ranking for acceptance into the broker system (see again box 2.2.5 in Chart 4.1).

Finally, the extent to which intervals are kept short will in large measure depend on the rate at which clients appear for acceptance into the system. It may reasonably be expected that, at least during the initial stages of operation, demand for services (or requests for acceptance) will build rather slowly. The fact that cumulative enrollments may be expected to build gradually gives the system manager during the early stages the option to control the threshold (and hence expenditures) without the necessity of dismissing those enrolled. Thus, the necessity to establish priorities or to severely curtail eligibility intervals may be eased a bit to the extent that clients appear gradually. *This leads to the suggestion that, political advantages aside, no major advertising campaign be launched in early stages.* A major influx of clients during any short interval could seriously hamper the integrity of the system and restrain the manager's ability to control (or balance) expenditures and demand.

RATIONALIZING THE REFERRAL FUNCTION

The referral mechanism is the heart of any social service program. It is here that the client is actually directed to—scheduled for—the treatments for which he is eligible. From the client's standpoint, then, this is the most significant component of the social service delivery system. From the managerial perspective, a method of scheduling or effecting client treatments should be developed subject to the following considerations:

1. Schedule the client only for treatments which are the responsibility of the HCDS (not those which are either gratuitous, or for which some other agency is properly responsible).

2. Do not neglect scheduling the client for treatments that are important and for which he is qualified.

3. Treatments should be delivered by adequate or qualified providers.

4. Do not have treatments provided at a cost which exceeds proper reimbursement levels.

Of these four objectives, items 1 and 4 pertain mainly to *efficiency* and are aimed at minimizing costs (resource expenditures) for any given level of service. Items 2 and 3, on the other hand, are concerned with *effectiveness,* with ensuring that the client gets adequate treatment for the full range of eligible conditions. The referral logic built into the HCDS system (and easily transferable to other social service contexts) attempts to realize these four objectives and to do so with a minimum administrative overhead. In this regard, consider master logic Chart 4.3.

Initially, it is possible that a referral request may be made either by a client himself, or on the client's behalf by some other social service agency (through the transfer mechanism), by a mainstream provider (e.g., a physician), or by HCDS staff themselves who, in following the client's history, uncover a need for treatment.* In either case, the first step in the referral system (box 4.0) is to examine a special kind of file which contains an index of all major medical conditions which might pertain to an HCDS client (the file 8.40 is detailed in the next chapter and is called the problem provider category index). This file has two purposes. First, it sets out a set of *diagnosis codes* so that medical conditions may be identified efficiently·and without ambiguity. The coding used may be that of the World Health Organization, Medicaid or any of a number of different indices prepared by members of the medical community (e.g., a study currently underway in New Jersey has developed a set of 383 medical diagnoses, which ostensibly exhausts the cases the normal medical provider might be expected to run up against). Next, it suggests

---

*The mechanism by which this latter can occur is outlined in the next chapter which examines the implications of the HCDS client tracking capabilities.

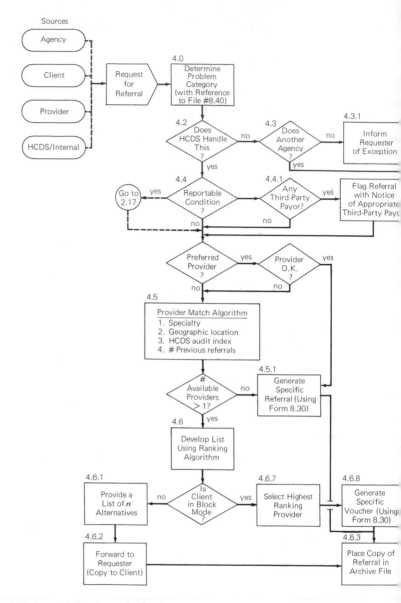

CHART 4.3 / Referral (client service) subsystem

what genre of medical provider is the appropriate agent for dealing with the various conditions—G.P.s, cardiovascular specialists, pediatric cardiologists, internists, etc. Thus, knowing the problem (diagnosis) code under which the referral is to be made also narrows down the range of providers to those appropriate

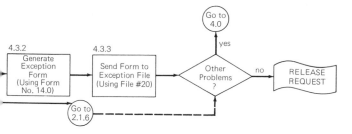

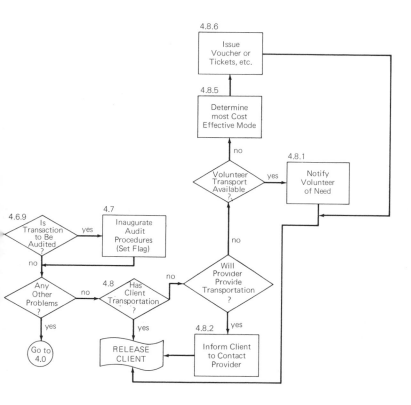

for dealing with the condition. The way to use this information and the manipulation of file 8.40 will be discussed in Chapter 5.

Given the recognition of the problem qua diagnosis code, it is then asked (box 4.2) whether this particular condition falls under the aegis of HCDS. If not, then the possibility that it is a condition for which some other agency is responsible is investigated. If so, send a transfer notice to that agency. If not, then an exception notice would be drafted, and the client (or his agent) informed that there is no existing coverage for the current complaint (box 4.3.3). A file of such exceptions is kept (file 20.0) and is periodically reviewed to keep abreast of those complaints for which referrals were unable to be made, given the current structure of general social service coverage. Should a number of clients be found to have requested a particular type of coverage, a move might be made to inaugurate a program or extend existing coverage into that area. This file 20.0 is thus a good source of information about lacunae in the aggregate social service system, pointing out what should be done that is impossible now because of categorical constraints or legislative oversight.

Moving on, box 4.4 asks if the condition for which the client is requesting treatment is *reportable*. If so, then inaugurate the procedures beginning with box 2.1.7 (Chart 4.1). A next question box (4.4.1.1) asks whether or not the client has third-party payor coverage for this particular condition. If so, then the eventual referral will indicate that HCDS is indemnified against reimbursement for this condition, and that the provider should be prepared to invoice the appropriate third-party agency (box 4.4.1.2). Following on, it is asked if the client has indicated a preferred provider, e.g., does he want to go to Dr. Jones or Dr. Smith or to some particular clinic or hospital. Upon such a request, a check is made to see if the provider is listed as approved by HCDS. If so, then a specific referral will be issued via box 4.5.1, and the notice of the referral placed in the client's file (box 4.6.3).

Obviously, no social service system need allow the client the option of preselecting his own provider. Where resources are particularly constrained or where certain providers tend to be overzealous in recruiting HCDS clients (e.g., where a Medicaid mill might exist), then the provision that allows clients to indicate a preferred provider may not be implemented. But where no exceptional conditions are present, it is generally in the service of client dignity to allow a client to express a preference and to honor it to the extent that the provider is approved, reputable and appropriate for the specific condition to be treated. However, where a significantly large proportion of the client population indicates a preference, the benefits from system-generated (rationalized) referrals is in part lost and so is an opportunity to control system effectiveness and efficiency. In short, there should not be a situation where the provider match algorithm (box 4.5) is sidestepped too often, for it is this algorithm that represents the referral "intelligence" of the HCDS.

Particularly, the algorithm seeks to refer the client to providers (or to some set of alternative providers) who appear to be *best* on the following dimensions:

1. Qualified to treat the particular condition for which the referral is to be made, that is, properly accredited specialists, or those who have the necessary facilities and equipment, etc.

2. Those as proximate as possible to the client (falling within a census tract area near to the client's residence to minimize travel time and costs, etc.)

3. The most cost-effective (or adequately cost-effective) for condition and geographic location concerned (see the next section)

4. Finally, assuming several providers emerge as well qualified on the basis of the above criteria, select that provider who has had the fewest referrals in the past so as to distribute referrals evenly among a large number of providers.

Thus, the referral logic seeks to isolate, from among the entire population of providers, that particular provider (or some limited set of providers) who promise to best serve the central system interests of efficiency (economy), effectiveness and client dignity.

Now, where there is only one provider who emerges as best qualified, then a specific referral is generated via box 4.5.1. Where several alternatives are specified (that is, where the algorithm has returned several providers as more or less equally attractive, given the selection criteria), then a list of *alternatives* might be generated and sent to the client (or the agent who initiated the request for referral). However, if the client is assigned to the block coverage modality, then that provider is selected who received the highest score from the algorithm, and a specific referral voucher is generated for the block provider selected (box 4.6.8). The point, again, is that clients assigned to the block modality are not expected to be able to make proper medical decisions on their own, and the system therefore exercises its responsibilities on the client's behalf.

Once a specific referral (or set of alternatives) is generated by the algorithm, the client is informed and a note as to the referral(s) is placed in his file for use at reimbursement time. Obviously, given a proper computer configuration—or an adequate clerical staff—referrals may be made in approximately *real time*, that is, with negligible delay. In many cases, the referral may be made during—or in response to—a telephone inquiry, meaning that the client would not have to appear in person to advantage himself of the referral mechanism. At any rate, with the referral(s) generated, the advisability of auditing this transaction is questioned, as per box 4.6.9. Some conditions—or perhaps some classes of providers—may be proper targets for a before-the-fact evaluation of their treatment procedures and plans, rather than waiting for a post hoc audit (of the type to be explained in the next chapter). In this case, the provider would be asked to inform HCDS authorities *before* he initiates a treatment procedure or as certain phases of treatment are completed.

To complete the referral logic, it is asked if there are any other problems for which the client needs an immediate referral. If so, return with the new condition back to box 4.0 and iterate the referral logic. If not, then check to see if the client is able to arrange transportation (if transportation subsidies are available). If so, he is released to get the treatment to which he has been referred; if not, transportation arrangements are made and the referral mechanism shut down for this case.

Most of the referral logic is relatively straightforward, then, and can be quite easily implemented in virtually all social service contexts (by programs referring clients to other than medical treatment). But the matter of assigning cost-effectiveness indices needs a bit more explanation. Therefore, this chapter on client logic will close with a brief explanation of how audit indices might be developed.

Developing Cost-Effectiveness Indices

As the reader is no doubt well aware, the matter of developing cost-effectiveness indices for professional providers is a matter of much sensitivity. Yet the integrity of the referral mechanism associated with any social service program demands that such assessments be made. From the standpoint of any health care delivery system—and the HCDS in particular—the support of medical providers will be had only to the extent that the cost effectiveness indices actually employed can be shown to be objective. In any case, sensitivities of professional providers aside, the enormous flow of social service funds into the mainstream medical community demands that some control be inaugurated.

There is, of course, the matter of the P.S.R.O. (Professional Services Review Organization) within the medical community. Under this system, providers are made at least partially accountable for their treatment decisions and assessed by their peers. This P.S.R.O. mechanism, however, does not completely resolve the matter of determining effectiveness criteria; it condemns only infrequently, and perhaps with reluctance. Moreover, it was designed more to meet the interests of private medical insurers than the needs of a broad-based delivery system. Therefore, the social service program must be prepared to establish its own cost-effectiveness criteria and to compute its own cost-effectiveness indices for individual providers.

The key to these indices recommended for the HCDS is this: to evaluate providers (a) to the extent they remain within the limits of acceptable medical practice in associating treatments with medical conditions, and (b) to the extent they remain within a reasonable cost range in charging for those treatments. This evaluative strategy really has the same effect as the PSRO scheme; it is the professional providers who set the standards. They are not imposed by laymen, nor are they imposed on localities by some central bureaucratic authority. In the HCDS scheme these standards are set not on the basis of case-by-case opinion taking (or peer review), but on the basis of certain broad statistical artifices. Here's how they work.

Initially, from previous exercise there already exists a set of established medical conditions which have been assigned an index number—the set of *diagnosis codes.* Now, from a number of different secondary sources, it is possible to generate data that express (either empirically or on the basis of subjective, judgmental evidence) the *relative frequency* with which certain conditions are likely to appear for treatment among the HCDS client population. What is being searched for is a set of M conditions which have a significantly high probability of exhausting the complaints of a majority of HCDS clients, where M is a *manageable* number.* Now, in order to arrive at the most favorable set

---

*Sufficient, however, to provide a representative *sample* of conditions against which effectiveness indices may be developed.

of M conditions from among the entire population of conditions which might conceivably occur for treatment, the data obtained is put into a frequency distribution as in Figure 4.4. The set of conditions (C) is then arrayed in order of expected frequency of occurrence; then the M enters as a threshold variable, and we have isolated the set of conditions most likely to exhaust the repertoire of client complaints, subject to the criterion of manageability dictated by M.

Keeping the list manageable becomes important because of the next step in the process. Particularly, we want to take these M most frequently encountered conditions and associate with each the treatments that are expected to be given in response to that condition. Again, we have the use of much secondary information of the type which relates diagnosis codes with subordinate treatment codes. Where no such information already exists, a survey of professional opinion may be inaugurated to get the correlations needed. At any rate, what we are after is a distribution as in Figure 4.5.

In this figure the horizontal axis arrays the set of all treatments that could possibly be associated with some particular medical condition, C_i, or, more generally, the set of all treatment codes (t) associated with any diagnosis code (D_i). It is possible to then define certain qualitative limits (or areas): area A would house those treatments that are deemed *essential* for the condition at hand; area B might contain a set of procedures that are optional, or which might be performed with limited frequency; finally, area C would contain procedures that are recommended only by a very small number of providers

FIGURE 4.4 / Frequency distribution of medical conditions

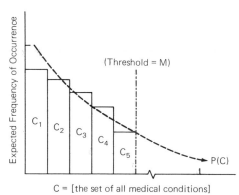

C = [the set of all medical conditions]

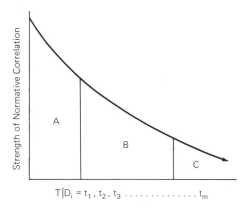

FIGURE 4.5 / Relationship between condition and treatments

and deemed gratuitous (or of doubtful utility) by a majority.

The point to this distribution is this: When a provider consistently uses treatments from area C—over several or many different conditions and patients—it can be assumed that he is so cautious and indifferent to costs that he is not an appropriate provider for HCDS referrals. Thus, the effectiveness index associated with a provider would actually be reduced by some incremental value every time he asks to be reimbursed for a treatment that, for the condition at hand, falls into the C region. In short, the cost-effectiveness index for a provider would be flagged for conditions of *overtreatment*. Thus, the provider match algorithm would reduce the probability of referrals being made to him in the future.

Perhaps even more serious are cases of *undertreatment*, where a provider fails to deliver a treatment that is deemed necessary or highly desirable by a majority of his professional counterparts. This situation would lead to a dilution of his effectiveness index, which would then reduce the probability of the algorithm generating referrals for him. Therefore, those providers who are most likely to be selected by the algorithm are those who consistently operate in the A range, such that the treatments they deliver are those generally accepted as being most appropriate for the condition at hand.

Thus, after several different provider-client contacts have been evaluated by HCDS audit personnel (and the mechanism for adjusting the cost-effectiveness index for a provider is built into the reimbursement logic explored in the next chapter),* both the overly cautious and the dilatory provider will have become improbable as providers, given the criteria under which the provider match algorithm operates.

As a final note, there is an optional procedure which might be followed in attempting to evaluate provider effectiveness. This would obviate the problem of generating a priori correlations between conditions (diagnosis codes) and treatments. Particularly, we could in the initial phases of the program begin with no presumptions about necessary, optional or gratuitous treatments. Rather, we could simply collect data as treatments are given and gradually build up an empirical (historical, a posteriori) set of correlations based on actual practice in the community. Once a sufficient number of provider contacts have been generated, then the kind of frequency distribution just illustrated would emerge naturally, and eventually be definitive enough to give us the control needed. In either case, we are still trying to maintain provider effectiveness within a range, eliminating both those that are inclined to do too little and those inclined to try to do too much. In short, the dilatory provider who does not perform the necessary treatments endangers the patient; the overly cautious provider, on the other hand, dilutes the system's economic base.

In practice, let's suggest that each new provider is given an effectiveness index that is initially equal to some maximal value, say 10. Now, on the basis of the client transactions that become apparent during the reimbursement process (where the provider specifies what he did for the patient relative to the condition or diagnosis), his index value would be reduced to reflect any instances of overtreatment or undertreatment, with more serious situations resulting in a relatively greater incremental reduction of the index. There is a final aspect to the effectiveness index which might be implemented in relatively well financed, more mature systems. This would employ a set of *conditional correlations* which would suggest the frequency with which a given diagnosis tends to lead to—or be associated with—subordinate diagnosis. Often a client going to a provider for one condition may be treated for some other condition. The conditional correlations would attempt to identify providers who consistently evolve subordinate conditions for treatment which are unlikely

*There is also a mechanism for re-referring clients who have been undertreated, for the client records—via reimbursement requests, etc.—will be flagged when a necessary treatment has been ignored.

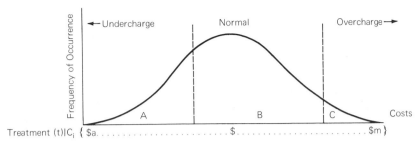

FIGURE 4.6 / Cost characteristics for treatments

given the statistical conditional correlations and who thus might be exploiting the system. It is well within our technology to provide this second-level control on effectiveness, though no such capability is as yet built into any existing program.

There is then the second aspect of the cost-effectiveness index—the direct-cost component. Note, first, that any provider would become less probable as the recipient of a referral as his cost index increases relative to any given effectiveness index. The logic for manipulating a provider's cost index would be very similar to that described for the effectiveness index. Particularly, we would want to develop—using either historical (empirical) or secondary (judgmental) data—a distribution as in Figure 4.6 for each unique medical treatment.*

In this figure we have assumed that we have placed a frequency distribution over a set of costs, for some specific treatment *(t)*, associated with some condition *(Cᵢ)*. For illustrative purposes, a normal distribution is used, but it could be of virtually any form. At any rate, a frequency distribution has been translated into three broadly defined qualitative domains: area A would contain distinctly favorable or lower-than-average costs for a given treatment; area B would expectedly contain the majority of cases; area C, on the other hand, would represent the domain characterized as overcharge, where costs are distinctly and significantly higher than average. Now, when a provider returns an invoice to the system asking for reimbursement for some treatment, we would note in which domain his cost falls. If, for example, his charge fell into area C, we would possibly want to increase his cost index by some increment, depending on the seriousness of the presumed overcharge. If, on the other hand,

*Or for some representative sample of *M* treatments.

his reimbursement request showed a charge falling into domain B, then his cost index would be left unchanged. Finally, if his charge falls into domain A, then we might want to actually reduce his cost index by some increment, thus noting him as a provider that is better than average in terms of cost factors. As his cost index falls relative to any given effectiveness index, the probability of his being selected for a referral increases, given the nature of the provider match algorithm.

In terms of operational significance, the cost index might thus be initiated at some arbitrary figure, say 10, just as was the effectiveness index. And the provider match algorithm would then act to give preference (priority) to providers to the extent that their cost-effectiveness ratio was low. That is, preferred providers would be those who consistently perform the appropriate treatments (no more or less) and do so at the most attractive fees. This, of course, is precisely what we want our referral subsystem to be able to do.

As a final note, it obviously becomes necessary to put *weights* on the several criteria the provider match algorithm takes into account. Thus, an unfavorable cost-effectiveness ratio may in part be offset by a provider's geographic proximity; or to avoid clusteration the algorithm may override a most favorable cost-effectiveness provider should he have received too many referrals in the past. At any rate, the precise codings and weightings are matters to be resolved by each system manager or by the operational staff of each social service program. All that was sought to be made clear in this section was that statistical distributions of a fairly obvious order could be used to engineer cost-effectiveness indices, and thereby allow programs to make use of the type of provider match algorithm just outlined.

It is now time to move on to the last chapter. There we shall be complementing this client's-eye view of a social service system with the perspective that becomes available when looking at a delivery system in terms of dollars and data.

NOTES AND REFERENCES

[1] For an expedient introduction to the concepts and implications of modern topology, see B.H. Arnold, *Intuitive Concepts in Elementary Topology* (Englewood Cliffs, N.J.: Prentice-Hall, 1962).

[2] For more on queuing theory and applications, see wither P.M. Morse, *Queues, Inventories and Maintenance: The Analysis of Operational*

Systems and Variable Demand and Supply (New York: John Wiley & Sons, 1958) or J.A. Panico, *Queuing Theory: A Study of Waiting Lines for Business, Economics and Science* (Englewood Cliffs, N.J.: Prentice-Hall, 1969).

[3] For most simulation exercises as they would arise in the social service sector, some sort of "programmed" simulation mechanism would be employed. SIMSCRIPT is a useful and important language, and is well introduced by Forrest Paul Wyman in *Simulation Modeling: A Guide to Using SIMSCRIPT* (New York: John Wiley & Sons, 1970).

[4] Some work has been done in the area of joint simulation-optimization. Particularly, see Karr, et al., *Simoptimization Research-Phase I,* No. 65-P2.0–1, C.A.C.I. (Santa Monica, CA: The RAND Corporation, 1965).

[5] For an analysis of methods for determining the "worth" of information—and hence the expected marginal product of information-generating instruments—see Chapter 5 of John W. Sutherland, *Systems: Analysis, Administration and Architecture* (New York: Van Nostrand Reinhold, 1975) or "Beyond System Engineering: The General System Theory Potential for Social Science System Analysis," *General Systems* (vol. XVIII, 1973).

[6] The elements of demand analysis are nicely discussed by Ferguson and Gould in Part I of *Microeconomic Theory* (Homewood, Ill.: Richard D. Irwin, Inc., 1975).

5

THE DOLLAR AND DATA DIMENSIONS

INTRODUCTION / In this fifth and final chapter we are going to narrow our focus much further than before. Particularly, we are going to scan in great detail the mechanisms of the HCDS that have the task of manipulating fiscal flows and information. The general procedures on the fiscal and data dimension should be typical of most social service delivery systems, but there are some aspects which promise to improve on current technology.

As for the fiscal aspect of social service operations, some time will be spent on the matter of regulating resource expenditures relative to budget conditions, as this is the major focus for basic system integrity. We shall also devote a few pages to outlining the reimbursement mechanism peculiar to the HCDS, and attempt to suggest how some of its characteristics may be transported to other social service systems.

Finally, we shall spend a great deal of time on the information processing functions of the HCDS prototype. These discussions are not simply for the information analyst or programmer, for the data dimension is the primary input to the decisions taken by the properly informed administrator or policy formulator.

THE FISCAL SUBSYSTEM

The fiscal system deals with the input and outgo—and internal manipulation—of dollars (or other primary resources). To a certain extent, this implies that the fiscal system represents the social service delivery system as seen through the eyes of the accountant. Most often, the accountant will tend to view the fiscal system not as an integral managerial instrument, but often as a relatively cut-and-dried response to the vast network of governmental accounting and reporting conventions. That is, when fiscal systems are established for the normal social service program, the attributes of the fiscal system are established, largely exogenously, in response to bureaucratic rules laid down by Medicaid, the Department of Agriculture or some other primary funding source. Like bureaucratically determined phenomena in general, these accounting and reporting conventions tend to be developed ad hoc, and thus may sometimes be both onerous and unwieldy. From the standpoint of the local social service manager, it must appear that virtually all federal and state agencies exist simply to invent new reporting requirements or to lay down new accounting conventions. As a result, the social service manager is deluged by a concatenation of data and reporting demands, at least some of which are of doubtful utility. Anyone who doubts this has only to look through the reporting requirements of, say, Medicaid. Not only are the reporting requirements enormously complex and demanding in terms of administrative overhead, but the demands—viewed in aggregate—are *irrational.* This irrationality stems at least partly from the following:

1. The aggregate of reporting requirements cannot be "ordered" into any *formal set* of data bases or informational items; that is, there are many requirements that have absolutely no logical connection with each other.

2. The aggregate of requirements contains many instances of redundancy, where the same data must appear in several different, uncorrelated locations.

3. From the standpoint of the information useful for policy or decision making, the aggregate of reporting requirements has significant lacunae or gaps. That is, despite the incredible degree of detail required, some information which has obvious utility cannot be retrieved at all from existing data bases.

4. Some data has no destination, that is, some pieces of information are not used anywhere at any time. They were asked for, it seems, within the context of bureaucratic "gamesmanship," simply to exert an apparent prerogative.

The telling point about the irrationality of the reporting requirements of the Medicaid-Medicare system (and of some other social service systems as well) may be summarized as follows: Firstly, the administrative overhead, relative to the utility or imputed value of the information collected, is dramatically high. In short, many of the reporting requirements are otiose and therefore preempt scarce resources to no real effect. Secondly, the reporting requirements do not always prevent abuse, exploitation, waste or corruption. Despite the masses of bureaucratic commandments and conventions, the auditing and control functions are at least in part abjectly *ineffective*.

It is interesting to note that governmental funding sources (the central bureaucracies) appear to be more rigorous in reporting requirements and fiscal controls than their counterparts in the private sector, (banks, investment houses, etc.). That is, when a commercial or private investor allocates resources to some company or enterprise, the relative reporting requirements (in terms of data demanded and information base conventions) are usually far less onerous than those laid down by the public bureaucracies. But, even though the reporting requirements are far less stringent and costly to execute, the average commercial investor has a better comprehension of the use and effect of its allocations than do any of the governmental bureaucracies. The reason for this is relatively straightforward. Private investment groups realize that once they lend funds, the analysis and review of reports (or the overview of the recipient enterprise) erodes the potential profit to be realized from that investment. Therefore, they have a direct interest in *minimizing the flow of information* from the enterprise receiving funds back to the enterprise which made the resources available. In short, the private investor always tries to maximize information leverage.[1] He wants as few man hours and computer hours as possible spent auditing activities of fund recipients and therefore always seeks the least information consistent with adequate control.

Now, just the opposite is true with most government or public bureaucracies. They gain their prestige and power largely through the number of individuals they employ. To maximize employees, they have an inherent tendency to maximize the flow of information from recipients of funds back to the bureaucratic investor qua conduit. Therefore, they spend a considerable amount of their time trying to figure out what new pieces of data may be demanded of programs out in the field. And because the generation of reporting demands is an ad hoc process, these reporting requirements are not always related to decision inputs. Indeed, some government allocative agencies have an apparently irresistible impulse to collect data, and often never get around to finding a use for it.

But there is another, more subtle reason for the dramatic lack of cost-effectiveness associated with most of the external reporting and control requirements imposed on social service agencies. Particularly, as already suggested, very few administrators in the social service sector have the technical or analytical skills to adequately manage the organizations with whose interests they are charged. As we move upward in some bureaucratic hierarchies, we find that managerial competence often tends to be diluted and not enhanced; more and more appointments to key positions are made on the basis of political affections or social relationships, and less and less emphasis is on demonstrable experience or expertise. Moreover, the senior positions in many of the central government's bureaucracies tend to be occupied by individuals whose education ceased many years ago, long before managerial technology matured. Therefore, paradoxically, managerial competence may often decrease as we move from the local to the state and finally to the federal level. With this decrease in technical competence comes the inevitable confusion about what information is really useful and about what legitimate (rationalized) reporting requirements should be. It is an axiom of information theory that the more competent the reviewing or control authority, the less information he requires in order to be able to make determinations about the health, propriety or cost-effectiveness of some program under his aegis.[2] In this light, then, the burdensome reporting requirements exacted of local programs by most central government bureaucracies (and to a lesser extent by their state-level counterparts) are the surest sign that somebody up there really does not know what he is doing.

But not all the excessive reporting requirements stem from the amateurishness of the reporting and review officials in the higher-order bureaucracies. Some of the more dubious requirements stem from the basic irrationality of the social service structure itself. Particularly, as explained earlier, the segmented, *categorical* basis of affecting and funding social service programs invites unnecessary administrative overhead. Each category of social service function belongs to one or perhaps several little bureaucratic agencies existing at the federal level (often in concert with a similarly mandated agency at the level of the several states). The bureaucratic impulse operates strongly even in these small categorical agencies, and much of their time is spent under the assumption that the more information they require of local programs, the better the prima facie case for their own competence as husbanders of the public resources. Therefore, whenever a local social service program agrees to accept categorical funds (and these are often the only source of input), they also agree to become the constant and passive victim of the reporting and control initiatives of the higher-order bureaucrats. As a consequence of their passivity, local system managers may often see the resources available for direct services constantly declining at the expense of more and more costly and time-consuming reporting and documentation requirements.

This is a good time to make some general assertions which will have some interest for the reader concerned with fiscal and system control. They are as follows:

1. No reporting requirement should be levied unless (and until) it is shown to be a direct, necessary and economically justified input into an essential managerial decision function. That is, no piece of information should be requested of local programs (or regional aggregates) unless its decision utility is a priori exhibited.

2. Bureaucratic agencies—at both the state and federal level—should be audited by a senior authority to note instances where reporting costs are excessive relative to base funding. The key basis of control of the bureaucratic conduits is the extent to which they are able to limit administrative overhead (i.e., transaction costs) relative to the flow of resources they make available.

3. As nearly as possible, external (state or federal, etc.) reporting requirements should parallel the reporting requirements necessary for local internal management of the social service program. That is, the structural correlation between local and exogenous (higher order) reporting demands should be as high as possible.

4. Finally, as much emphasis as possible should be placed on developing reporting (and hence information and fiscal con-

trol) systems that has the lowest degree of information redundancy in terms of storage and generation requirements, and which thus make each datum serve the greatest number of decision input functions.

These are some of the propositions that should be considered in the design of any fiscal and reporting subsystem in any social service context.[3] It would be fatuous to suggest that providers of resources—government bureaucracies or otherwise—do not have both a right and responsibility to demand an accounting of the uses to which funds were put. Therefore, the above suggestions should not be construed as an argument against reporting requirements. Rather, they simply ask that reporting procedures and demands be reconsidered in the light of the funding sources' broader responsibility: to rationalize information demands and constrain them to those that have a demonstrably positive cost-effectiveness index (which means they have a strongly positive decision utility).[4] It is in light of this assertion that we may consider the fiscal subsystem built into the HCDS prototype, illustrated as Figure 5.1.

As may be seen from this figure, the prototypical fiscal subsystem should be prepared to accept—and manipulate—both categorical and undedicated (unallocated) funds or resources. For the case of the HCDS there are thus five different funding sources that are anticipated, and each will be handled within what is called a *flexible fiscal logic*. The flexible fiscal logic is probably unique to the social service sector, and it reflects a simple suggestion: to the usual social service program, *all dollars are not the same.* As suggested earlier, then, the social service manager—qua fiduciary functionary—has problems that his governmental and commercial counterparts may not face.

Particularly, when speaking of the fiscal responsibilities of the average business manager or even the usual public administrator, his duties become intelligible in terms of a proposition offered earlier in these pages: every allocation of resources should carry a *zero-opportunity cost* (meaning, again, that no better opportunity existed for the use of the funds than that to which they were actually put). An important qualification, however, is that the funds are not a priori dedicated to some specific function or earmarked for some predetermined destination. This does not hold true in the social service sector simply because of

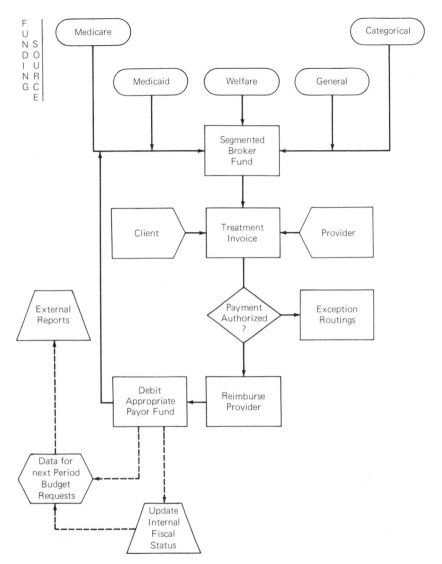

Medicare

Categorical

Medicaid

Welfare

General

Segmented
Broker
Fund

Client

Treatment
Invoice

Provider

External
Reports

Payment
Authorized
?

Exception
Routings

Debit
Appropriate
Payor Fund

Reimburse
Provider

Data for
next Period
Budget
Requests

Update
Internal
Fiscal
Status

FIGURE 5.1 / A prototypical fiscal subsystem

the categorical constraints imposed by so many funding sources. In the case of health care systems, for example, Medicaid, Medicare, Welfare and other categorical funds are all somewhat independent of the local system manager's discretion or judgment. What is called "general" funds in Figure 5.1 are a priori unallocated with this proviso: that they are not spent to support a function to which one or another of the categorical funding sources is dedicated (unless demand for services exceeds the supply of categorical funds).

The zero-opportunity cost proposition thus takes on a special implication in the social service context. For each of the categorical funds (or budgets) carries an *artificial* zero-opportunity cost; that is, they are either used for the purposes predetermined, or not used at all. This means that the local system manager cannot be held accountable for the wisdom and efficiency with which he uses his aggregate funds, for one dollar is not like another. Rather, he can only be responsible for the allocation decisions he makes *within* each of the categories, and not *among* them. He cannot, in short, transfer funds from one categorical budget to another (at least not with impunity). In many cases, he may be absolved from responsibility for the allocation of resources within categories, for often these are capitation budgets (or sliding budgets) that are to be spent on demand and not on the basis of discretion.*

Again, then, the categorical constraints that dominate the existing social service sector severely limit the fiscal authority of the local manager and therefore limit the extent to which he can legitimately be held accountable. Rather, it is the central funding authorities themselves (e.g., the Medicaid complex) who are to be held accountable for the integrity of the aggregate of resource expenditures. Because of the enormous domain and great swathes of time over which these central systems hold aegis, holding them accountable implies a level of auditing technology that we simply do not possess or which would command enormous resources. Therefore, one of the key postulates of modern system science comes here to haunt us: *if a system can grow big enough and be organized in a sufficiently incoherent and obtuse way, it becomes immune to accountability.* The same cannot be said of local programs or even well-defined regional complexes. Therefore, on the basis of accountability alone, the argument goes strongly against a centralized social service structure and begs for decentralization.†

The direct implication of the categorical modality is that local programs need to maintain a *segmented* funding base. Simply, the fiscal subsystem of the typical social service program is

*This point will be amplified in the next section.
†Again, however, the critical qualifier will be the extent to which managerial technology increases in sophistication.

required to maintain separate accounts (and many times separate escrow type holdings) of the resources contributed by the several categorical funding sources. A dollar from Medicaid, for example, cannot be confused or merged with a dollar from the welfare fund or with a dollar from the general operating account. Continuing, this implies at least two subordinate capabilities for the fiscal subsystem: (a) it must be able to distinguish the clients that are uniquely the responsibility of one or another of the several payors (funding sources); (b) it must be able to distinguish transactions or individual treatments that are to be charged to any particular payor; (c) and given these broad responsibilities, the fiscal subsystem must also be.able to do the following:

> Resolve instances where a single client (or a single client-treatment combination) could be allocated to two or more payors, or where there are discontinuities in a transaction, e.g., where part of an ostensibly infrangible treatment is to be charged to one payor, and part to another.

> It must be able to immediately determine residual responsibilities; in short, to determine those cases where no payor may be charged, which implies that the transaction is to be debited to the "general" or unrestricted account.

> It must be able to give real-time (or effectively so) reports on the status of the various payor accounts, such that we can comprehend the rate of exhaustion of that account relative to any dollar, temporal or contact limitations.

In some cases, these demands exceed the capabilities of the existing fiscal subsystem. In such a case, the local authority simply does not know what he has spent where, what remains to be spent or for what charges he is reimbursable from the central funding authorities. For example, it has been estimated that New York City fails to receive many millions of dollars a year to which it is legitimately entitled simply because it cannot coordinate its expenditure patterns with the provisions of its many categorical funding sources.

The mechanical requirement, then, is that the fiscal subsystem have a central document to which it responds, and that that document be processable so as to determine third-party or categorical payor responsibility and also any residuals chargeable to the general account. This document, in the HCDS system, is the *treatment invoice,* and will be examined in the next section. On Figure 5.1, then, once the treatment invoice has been received, the reimbursement cycle is kicked off. Particularly, there are various procedures to go through to see if the treatment for which reimbursement was requested is indeed authorized and, if not, how exceptions are to be handled (these functions will be discussed shortly in concert with master logic Chart 5.1). If payment is authorized, then the fiscal system routes the payment to the provider and then debits the appropriate payor fund (e.g., the categorical fund or budget segment which was deemed responsible for the client and/or treatment transaction just reimbursed). When a fund has been debited, the fact of the debit now has an impact on the central internal management variable, the rate of expenditure of that fund. Finally, the fact of an expenditure for a certain client-treatment combination implies a probability that that transaction will be repeated in the future and therefore generates a datum that should be available to the planning function (responsible for projecting budget requirements—by category—for a future funding period).

Most of these operations are relatively simple and straightforward. But a better understanding of the fiscal functions can be gained by looking at the reimbursement process in a bit more detail. In this regard, then, consider master logic Chart 5.1.

The first of the reimbursement functions (box 6.0) requires *sorting* what is called the *transaction file* (file 12.10) by *provider.* (The composition of the transaction file and the description of the transaction invoices it contains will be described in the final section of this chapter.) In the broadest terms, however, a *transaction* is the invoice that is returned by a provider which specifies what treatment for what condition was done to what client at what cost. The heart of any social service delivery system—irrespective of whether or not it uses mainstream providers—is always the type of transaction mentioned here. It is both the primary evidence of the system's activities and the datum of principal fiscal significance. The next step is to select a particular provider and begin the processing of all the invoices (transactions) to which he is a party. A particular invoice is then selected (box 6.2) and then considered in light of the cost-effectiveness indices discussed as part of the referral logic. Initially, we want to be able to evaluate the extent to which the treatments given by the provider were congruent with the diagnosis (medical problem) of the client. Recall, from the last section of Chapter 4, that we had developed correlations between conditions and treatments, such that for any given condition (diagnosis) some treatments were deemed necessary, while others were evaluated as optional and some as gratuitous (unlikely to be useful). Given the correlations, we then examine each condition-treatment set for each transaction and search for situations of overtreatment or undertreatment. If there is an indication of overtreatment, then the effectiveness index for that provider is debited according to procedures outlined in Chapter 4. Conditions of undertreatment calls for another level of concern. Not only is the provider's effectiveness index reduced, but we want to be alert to the possibility that the neglected treatments should be performed. In short, where analysis of the treatment invoice (the transaction) indicates that some expectedly necessary treatment was foregone, we would want to refer the client for that treatment. This is the link back to the referral subsystem, indicated on the logic chart following box 6.3.2.5.

It is also necessary to evaluate the current transaction with respect to the cost component of the provider's cost-effectiveness index. This is done at box 6.3.3, where the compensation the provider demands for the treatment is compared against some standardized cost distribution (of the type illustrated in Chapter 4.). As with the effectiveness index, deviations from the tolerable limits are noted and result in an incremental increase (for instances of overcharge) or decrease (associated with a lower-than-average cost) of the provider's cost index. A check is then made to see if the client is a copayor (box 6.3.6). If so, and if we are keeping copayor accounts rather than simply paying the provider the residual, then an account receivable is established for the client (or we just increment his outstanding account). Generally, a delivery system may expect to get more enthusiastic support from mainstream providers by having the system, rather than the provider, seek recompense from copayors. This saves the provider from extra clerical work and also has another very legitimate advantage. Should a copayor fail to meet his obligations, his coverage may be curtailed. In this regard, what the system has spent for him and failed to recover becomes a *credit* against what would have been spent had he remained enrolled. Therefore, there may be no *real loss* to the social service system should a client not clear his account receivable. It is important,

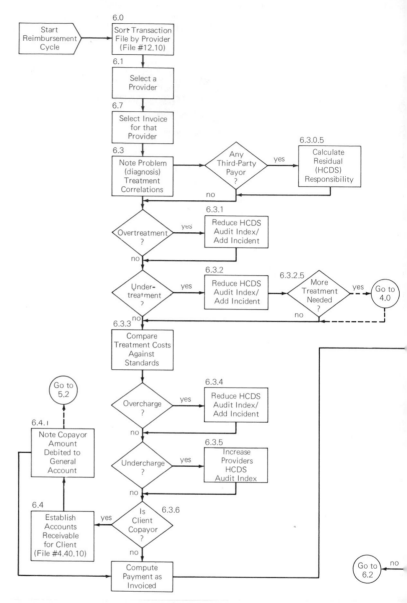

CHART 5.1 / Reimbursement functions and fiscal processes

however, that no account receivable remain unserviced for a long period of time or be allowed to grow too large. So the referral subsystem should recognize delinquent cases and refuse to authorize treatments after some time or dollar threshold has been exceeded.

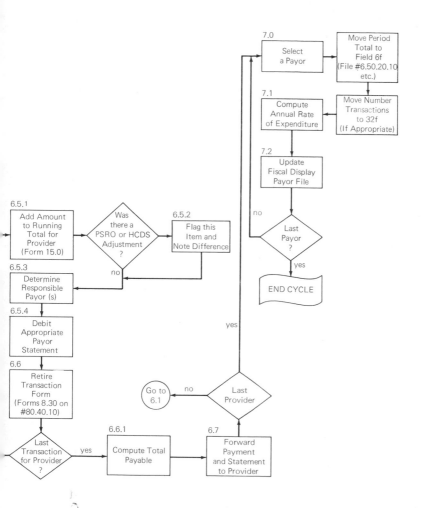

Moving on to box 6.5.1, we then credit the provider's account for the amount of the invoice.* For some transactions, ~~you~~ may be what is called a P.S.R.O. flag; this indicates that the

---

*Less any third-party payor amounts determined in box 6.3.0.5.

P.S.R.O. mechanism was asked to review this particular provider-client contact and report back on the legitimacy of the treatments and/or costs. Where a P.S.R.O. review is outstanding, then the credit to the provider's account payable is held in abeyance, pending resolution. In other cases, the HCDS will have entered an adjustment, especially when the amount of reimbursement requested is way out of line with the standardized cost distribution. Occasionally, the P.S.R.O. process will recommend a reduction to the amount invoiced. In either of these cases, then, the provider is credited with an adjusted amount (box 6.5.2), and the reason for the adjustment indicated on the provider statement. Next, as is consistent with the logic developed with respect to Figure 5.1, we determine the payor (funding source) responsible for this transaction and debit its account (box 6.5.4). Finally, the transaction is retired to an archive or holding file, where the information will subsequently be made available for periodic reporting purposes.

The remaining steps in the normal reimbursement cycle are obvious: simply repeat the logic (beginning with box 6.2) for any remaining transactions involving that provider and finally for all remaining providers. At the conclusion the statements and reimbursement checks are forwarded to the various providers. There is a last set of functions to be performed at the conclusion of the reimbursement cycle (or perhaps after some number of cycles have been completed, if the reimbursement frequency is high, i.e., weekly or bimonthly). This involves the matter of trying to determine the rates of expenditure for the various payor (or segmented) accounts.

The Internal Control Logic

When speaking of internal control, the concern is with the fiscal integrity of the social service program. Relatively familiar budget control procedures can be employed here, providing the peculiarities associated with the *segmented* funding base are recognized.[5] Fiscal integrity is the concern of tasks 7.0–7. on Chart 5.1, and we may lend them a little substance in these xt several pages.

The basic internal control problem is the matter of coor-dinating rates of resource expenditure with respect to budget limits for each of the various segmented funding bases (payors). The three subordinate variables in the controls process are: (a) the size of the various segments (the budget limit); (b) the size of the client subpopulations associated with the funding segment in question; and (c) the level of service the system will deliver to clients within the given funding category. Level of service has two different dimensions. Firstly, the level of service may imply the *breadth* of coverage (the number of different conditions or casualty states the system will seek to correct or treat). Secondly, it may imply the *intensity* of treatment provided any given condition. In the health care sector, for example, a program might increase the level of service by expanding its coverage to include, say, psychiatric or maternal disorders in addition to acute injury or disease conditions, etc. Intensity of service might be increased when we elect to allow more dollars per treatment (e.g., permit the use of expensive therapies, drugs or equipment; allow the procurement of relatively more expensive prosthetic devices).

At any rate, the way in which a particular social service system resolves these three variables is the signal of the competence of its local management. More specifically, the system manager may adjust client population by raising or lowering the qualification threshold (as explained in Chapter 4). He may adjust level of service by raising or lowering breadth and/or intensity of coverage. He may, of course, adjust both client population and service level simultaneously, always with respect to the key control variable: rate of resource expenditure relative to resource (budget) limits. Given the segmented funding base of the usual social service program, these adjustments must be made for each of the several different categorical payors. Clients (and/or conditions and treatments) must be viewed as the responsibility of particular payors, thus complicating fiscal procedures. But there are other complications as well.

For example, different payors may employ either of two different budget modalities. *Block* budgets supply the program with a given quantum of dollars (or other resources) relative to some time frame. That is, block budgets may refer to a particular interval of support (say a fiscal year). The *capitated* budget, on the other hand, is calculated relative to a given client population, or relative to some expected number of service contacts costed at a standard rate. Therefore, rates of expenditures, under the two types of budgets, are calculated with respect to different bases: (a) rate of expenditure relative to the proportion of the fiscal year exhausted (for block budgets); (b) rate of expenditure relative to the proportion of the capitation population served.

The system manager, when dealing with a block payor, wants to exhaust his budget by the end of the fiscal year without imposing too much disparity of service levels at different times. The manager husbanding a capitation budget also wants to exhaust his resources at the end of some period, but also must make sure that a certain number of clients were served in the process.

If there are two different budgeting modalities, there are three different types of fund transfer mechanisms which must be considered. These further complicate the fiscal portrait of the social service sector. Specifically, block or capitation budgets may be delivered to the program in three ways: via *drawing* accounts, as *escrow* accounts or as *demand* accounts. The distinctions between these different types of accounts are important. Under the drawing account, a program is given a certain fixed budget against which it can draw as expenses are incurred, but the aggregate allocation cannot be exceeded for any given interval (thus even capitated budgets may have a tacit time dimension, e.g., a certain number of service contacts are to be funded with respect to some fiscal year). An escrow account also has a fixed aggregate limit, but represents a one-shot transfer of the entire budget amount to the local program at the initiation of the fiscal year. Finally, the demand account usually finds the social service program being given a certain "seed capital" fund to initiate the fiscal year; thereafter, the local program is simply reimbursed for expenditures as they are made with no fixed budget limit being set. In short, from the standpoint of the local system manager, there is an unlimited budget at his disposal.

The demand account, of course, imposes the fewest problems on the local system manager, but also invites the type of profligacy noted with respect to the Medicaid program and many of the income-support or antipoverty programs. The escrow account and the demand account both make the local manager responsible for staying within the aggregate budget limits over the interval in question. The drawing account allows the central funding authority to "stage" its fund transfers and perhaps adjust them in response to quarterly or other periodic rates of utilization. In short, the drawing account enables exogenous control *within* fiscal periods, where the escrow account does not. The transaction costs, however, are higher with the demand account than with the escrow mechanism; because the reporting mechanisms which are in force for existing demand mechanisms are largely inadequate, the net effect of the two modalities is the same—the system manager has the implicit responsibility for completely using the aggregate of funds available to the program, even when some portion of the allocated resources could be used to better effect elsewhere. But, again, the artificial zero-opportunity cost associated with categorical funding in general does not argue for individual system managers looking for better uses for resources *outside* their own system.[6] Therefore, the only real differentiating variable is transaction costs, and on this basis alone the escrow account is to be generally preferred to the demand account.

The system manager under either the escrow or demand accounting mechanism and facing either a block or capitated budget has essentially the same problem. His task is to effect a more or less *symmetrical* rate of expenditure across a fiscal interval or across a client (contact) population.* But this implies that he is able to predict—with some degree of accuracy—the schedule at which clients will appear for service and the average expenditure per contact or treatment. Knowing this, he could then develop a resource expenditure algorithm that would exhaust the budget by the interval in question or over the required number of clients or contacts. The development of such an algorithm presumes that the social service program has a long and stable history on which it can predicate such projections. For new programs, or for programs operating in environments that are subject to change, any projective or distribution algorithms will entail a significant error component. For this reason, the method of determining proper rates of expenditure would be different for the mature, stable program than for the unprecedented or protean program.

In general, however, for the moment considering only block budgets, the system manager may know that the appropriate service level (breadth and intensity of coverage) and/or the selectivity of the qualification threshold (determining level of enrollments) must be a function of:

1. the total block budget available for the fiscal year

2. the proportion of the aggregate budget already expended by some time (t_i)

*Such, that is, that the "level" of benefits provided the stream of clients is more or less symmetrical across a period.

3. the projected demand for services (or enrollment) for the remaining portion of the fiscal year $(t_i \rightarrow t_n)$

Now the problem of the system manager operating an immature social service program—or one resident in a rapidly changing milieu—becomes evident. He cannot make adequately accurate predictions about the schedule of demand for services throughout the fiscal period $(t_o \rightarrow t_n)$. Therefore, he must operate under the assumption that the first year or so of operations (or the current period) will provide him with a learning experience for subsequent periods. In this regard, he must make a *deduction* about the schedule at which clients will appear for service. Now, there are basically three different deductive referents the system manager might want to employ in establishing a base demand schedule, each of which will be appropriate as indicated:

1. Where he has no information at all about demand, he might assume that demand (and average cost of contact) will be distributed *symmetrically* throughout the period in question . . . that demand will be constant.

2. Where the type of coverage provided is essentially new— and where program promotion has been contained—he might want to impose a *learning-curve* demand function, one where demand accelerates through time.

3. For some programs, especially those designed to treat acute conditions on a one-shot basis, the manager might want to use a *skewed* distribution, suggesting that demand for services will begin strong and then gradually trail off.

In some cases, the manager may be able to inject some sort of empirical constraints which will serve to generate a referent distribution.[7] For example, an income-maintenance program may expect to have demand geared in some way to seasonal factors, especially in rural areas or in regions where employment itself is seasonal (as with primary industries like logging, fishing, fruit picking, etc.).[8] But there is a strategy already mentioned which can ease the prediction problem somewhat: if the system manager is allowed to maintain control over the access modalities employed and the qualification threshold, then he can at least partially control the demand for services. And what can be controlled, does not need to be predicted. Again, then, there is an argument for a context-responsive delivery system.

To now indicate something of the technique of managing expenditure rates, let's examine the situation where the manager is concerned with husbanding some block budget under an escrow account for an immature program.

Let: $t_\phi - t_n$ = the total fiscal period, divided into 52 segments, $(t_k = 1, 2 \ldots \ldots 52)$

b_j = the total budget (in \$) allocated by payor j for the period $t_\phi - t_n$

b_c = the amount of payor j's budget already expended by some time—t_k (a cumulative figure)

b_k = the incremental average expenditure (b_c / k)

If it can be assumed that demand for services remains constant throughout the year, such that each $b_k = b_j$, then the following ratio may be used as the key decision point: $\$b_j / b_c = n/k$. That is, the proportion of payor j's budget expanded at any given time should be related directly to the proportion of the fiscal period expired.

Now, if the ideal rate of expenditure were graphed, it would fall along the 45° curve of a diagram relating cumulative expenditures ($\$b_c$) and cumulative time ($t_k$). Suppose, just for example, that b_j were to equal \$100,000, then we would get Figure 5.2. For control purposes, for each noncapitation payor this figure would represent the *central reference*, the ideal situation. At any given time, then, the actual b_c and actual t_k should be compared to this reference, and the amplitude of deviation from the reference noted.

Suppose, for example, that we had an actual curve that

FIGURE 5.2 / The referent expenditure function

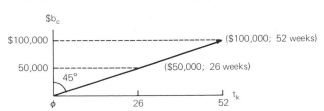

looked like the one in Figure 5.3, relative to the reference at the week ending at $k = 26$. It would be clear that we are actually overspending. If this rate were allowed to continue, the total budget would be completely exhausted by about the 35th week (leaving no budget for the remaining 17 weeks), as may be determined by examining the basic *control ratio* below:

$$\frac{b_i}{b_c} = \frac{N}{k} = \frac{\$100,000}{75,000} = \frac{X}{26}, X = @\,35$$

The nature of the corrective action which must be taken due to early profligacy is clear: the rate of expenditure for the periods $k = 27 \rightarrow 52$ must be radically reduced. The proper rate would be determined as b_k (the average weekly expenditure), where: $b_k = (b_j - b_c)/(n - k)$ For the current example, this would yield

$$b_k = \frac{(\$100,000 - 75,000)}{(52-26)} = \frac{\$25,000}{26} = \$962$$

The rate of expenditure for the period $k = 1 \rightarrow 26$ was

$$b_c = \frac{\$75,000}{26} = \$2885$$

Therefore, the rate of expenditure for the period $k = 27 \rightarrow 52$ must drop by a factor of 2885/962 or @ 3.0; this reduction to the remaining period of the fiscal year should yield a correction curve which will allow arrival at the target figure of $100,000 in 26 weeks, as in Figure 5.4. One thing which should be very clear: the earlier one identifies an improperly accelerated rate of expenditure, the less dramatic the corrections which must be

FIGURE 5.3 / Overexpenditure of a fixed budget

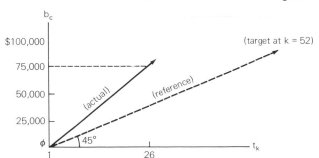

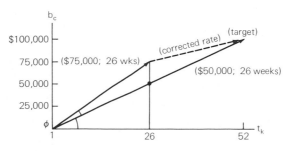

FIGURE 5.4 / Corrected expenditure trajectory

made. Hence, recommended is the production of an *actual rate of expenditure report* for each noncapitation payor every week or geared to frequency of use.

Essentially, the same logic would hold were it shown by the weekly update of the central chart that we were expending at too low a rate, for example, Figure 5.5. The central ratio at week $k = 13$ would now read something like this: if $100,000/(b_c) = 52/13$, then b_c should $= 25,000$; but, the actual ratio is $100,000/18,500 = 52/13$ which means that we are underspending the funds provided by payor j (e.g., we should have, according to the reference ratio, expended $18,500 by the middle of the 9th week rather than by the end of the 13th). So an upward correction must be made for the remaining periods, according to the following:

$$\frac{(b_j - b_c)}{(n-k)} = b_k = \frac{\$81,500}{39} = @ \$2,100$$

Previous expenditure was at the rate of $b_k = \$1,428$, so we were underspending by a factor of $2115/1428 = 1.5$.

FIGURE 5.5 / Underexpenditure of a fixed budget

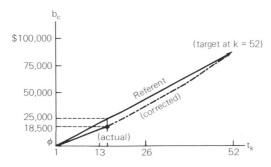

Now, the presumption that services will be demanded at an approximately equal rate during the fiscal year (e.g., such that $b_1 = b_2 = b_3 = \ldots b_{52}$) is allowable only for the first year of a system's operations (and not even then if there is information which would suggest some other pattern). For after a year is past, the rate at which services were demanded during the first year becomes a crude reference for the prediction of the rate at which demands will be made on the system during subsequent years.

The central logic expressed above for the first year (where a constant demand for services is assumed throughout each week) may, in subsequent years, be elaborated quite simply.

First, take total expenditures for the first broker year and take the average: $\bar{b} = (b_c / k)t_n$. This $\bar{b}$ then becomes the *standard* expenditure, $S_j, t-1$. Now, each week of the previous year will be examined with respect to this reference, such that an index number results: if S_j (1975) = $1,000 (e.g. $52,000 ÷ 52 weeks) and $b_j,1$ (1975) = $1,250, then expected $\bar{b}j,1$ (1976) will be 1.250.

Now, the 1976 budget for payor j would probably be different than that for the previous year (1975), so the standard expenditure for 1976 would be some new figure, $S_{j,t}$ (e.g., $104,000). Now, the expected $\bar{b}j$, 1 for any week would be the index number from 1975 times the S_j,t value.* Consider the following examples:

1975 (First Year)	*1976 (Second Year)*
b_j = \$52,000	b_j = \$104,000
k = 52 weeks	N = 52 weeks
S_j = \$1,000	S_j = \$2,000
if: $b_{j,1}$ = 800	then: Exp. $\bar{b}_{j,1}$ = 1,600
	(e.g., .8 × \$2,000)
$b_j,52$ = \$1,800	then: Exp. $b_j,52$ = \$3,600
	(e.g., 1.8 × \$2,000)

Now, the expected values of all $b_{j,k}$ for 1976 would be used to generate a second reference curve in addition to the 45° curve introduced previously (see Figure 5.6).

*Note that this is an extremely crude process and would be used only when seasonal, cyclical, etc., information were not available to support a proper time-series analysis.

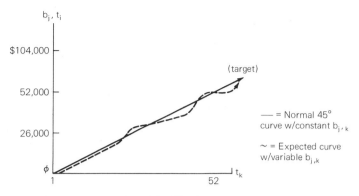

FIGURE 5.6 / Modified referent function

If, now, a condition of overexpenditure or underexpenditure is identified at some point in time, corrections should be made using the expected values of demands for the remaining periods ($k = i$–52), where a revised (upward or downward) S_j would be calculated and multiplied by the appropriate index numbers to yield week-by-week expenditures which would have the highest probability of meeting the operational target (e.g., $104,000 at 52 weeks).

Thus, the actual expenditures during year (t–1) serve to discipline our predictions about year (t) and improve our control capabilities. Moreover, the expected values may be presumed to become more precise as there are successively more years from which actual data may be collected. For the first year, however, a constant rate of expenditure may be presumed and the charts for each payor constructed and analyzed accordingly.

The logic for the payors whose contribution to the broker fund are based on per capita rates is essentially the same as for block payors except for the following:

1. The t_n and t_i of the previous sections now becomes replaced by:
   M = total number of clients or service contracts budgeted for the current fiscal year
   N_i = the cumulative total of clients served or the cumulative number of service contacts to some period ending at t_i

N_k = the number of clients served or contacts executed during some period k ($k = 1$–52)

2. The relevant control ratio now becomes $b_j / b_c = M / N_i$

This would generate a curve exactly like that of Figure 5.2 and yield essentially the same central logic: expenditure periods and expenditure rates relative to those periods have now simply been displaced by numbers of capitated clients (or service contacts) and the relevant rate of expenditure is now referenced against these numbers rather than time. Thus, the control of capitation expenditures is to ensure that the payor's number expectations are met and that his budget is used fully.

Finally, the system manager will put the rates of expenditure relative to numbers (for capitation payors) and relative to time (for noncapitation payors) to work in controlling the three key parameters of the broker system:

1. Enrollment threshold

2. Cost to clients/(copayor provisions)

3. Breadth of services offered (or intensity of treatments)

Given the logic already developed, the nature of the decisions which must be made are evident.

CONDITION I: *Underexpenditure of Resources from Noncapitation Payor*

1. Decrease enrollment thresholds so that new clients may be admitted to the system at a rate which will be expected to raise the average (or expected) $b_{j,k}$ to the required rate, such that the fiscal target will be met at $k = 52$. This would involve reducing the barriers to enrollment (e.g., raising minimum salary levels) for the category of client supported by payor j.

2. Leave enrollment threshold constant, but decrease client copayment proportion, or extend services of certain sorts to the nonindigent population (e.g., allow nonclients to gain access to the referral subsystem).

3. Extend the breadth of coverage to clients (e.g., allow them to have general medical screenings or dental care paid for).

CONDITION II: *Overexpenditure of Resources from Noncapitation Payors*

1. Raise enrollment threshold to curtail client entrance.

2. Set a higher copayment rate or set a ceiling coverage limit per client; restrict access to system services.

3. Cut back the range of coverage allowed clients.

CONDITION III: *Underexpenditure of Resources from a Capitation Payor*

1. If the condition results from too few clients for the category stipulated relative to the number budgeted, then two remedies are available (assuming that budgeted funds should never be returned to a payor):
   a. Recruit new clients for this category.
   b. Shift to the extent permissible any general HCDS clients to the category (or provide the categorical care under the budget without reference to any other requirements of the category).

2. Transfer any excess funds resulting from an average payment below the expected to the general broker account for use by noncapitated clients, providing that the number of required contacts or clients (the M figure) can be obtained at $k = 52$. This could be done by charging a higher administrative overhead rate, and thus moving the funds to another account while still meeting the raw capitation requirements. Where reporting requirements permit, funds may be transferred without any exotic "laundering" through accounts.

CONDITION IV: *Overexpenditure of Resources from a Capitation Payor*

1. Apply for an increase in the budget showing the higher than expected (or budgeted) average cost per contact or client.

2. Apply for a reduction in M.

3. Decrease, to the extent possible, the level of care associated with each client or service contact, so that the average cost will approach b_j / M.

As a final note to the matter of fiscal logic, it should be clear that the procedures introduced here are very simple. Where the skills and administrative resources permit, the calculus-based functions could be substituted for the simple algebraic operations illustrated. Moreover, rather than making adjustment to expenditure trajectories by "eye" (or with reference to simple indices), the technologically sophisticated program manager will have a simulation model at his disposal which, when programmed with expenditure, population, service and qualification parameters or projections, will generate an approximately optimal *simultaneous* solution. The expense associated with such models would, however, be justified only for fairly large programs (or perhaps were the program to be developed as a joint effort by several or many smaller agencies, each paying a pro rata share of the development and implementation costs).

Given this rather brief outline of fiscal and internal control procedures, we are now in a position to conclude our inquiries into social service management. So, in the next and final section of this book, we turn to an analysis of the information processing aspects of the HCDS prototype and thus look at the social service delivery system from the data dimension.

DATA PROCESSING LOGIC

Now begins the discussion of the last aspect of the prototypical social service delivery system, the *data dimension.* What is done here will be of interest primarily to system analysts and other technical personnel concerned with the clerical functions required of a delivery system. In the course of discussing the data processing functions, there will also be an opportunity to reinforce some of the arguments raised earlier about client and fiscal logics.

Without going into great detail, the performance index for the data processing subsystem is *leverage.* * That is, we want to

---

*In practice, the leverage of an information base would be assessed by translating the data bases into "trees" or some sort of network. Each datum would represent a node, and each manipulation function would be a branch. Leverage is then reflected in the number of branches associated with any (and ultimately all) nodes, and to the extent that each node is unique (and each manipulation function an iterable, proper subprogram).

serve the essential decision and reporting functions but at the same time produce the least amount of information, minimize storage requirements and do as little data manipulation as possible. The information dimension of the HCDS was designed with this objective in mind, and therefore might serve as a useful practical reference for system analysts engaged in other social service contexts.

What will be done in the next few pages is to indicate the way in which data moves through the HCDS, and also to show the relationship between the data processing dimension and the substantive system functions (access, referral, etc.). In this regard, consider Figure 5.7.

The data processing subsystem is initiated at the point where a potential client is accessed (either through one of the normal access modalities or through the contingency enrollment option). This results in the preparation of a client profile for the client. A client profile is divided into three segments:

Part I   (Form 5.20.10) contains the personal and qualification data for the client, developed as a part of the enrollment and socioeconomic (behavioral) screening process.

Part II   (Form 5.20.20) is the administrative record of the client, and shows what referrals have been made, what treatments he has received, etc., and what reimbursements have been made on his behalf.

Part III   (Form 5.20.30) contains the client's essential medical record, and thus serves as a coded abstract of his health problems or of other conditions that providers might wish to be informed of before making a diagnosis or delivering a treatment.

For obvious reasons of security, it is important that parts II and III not be easily connected with Part I. In short, we want to be sure that unauthorized people will not be able to connect a particular medical record or schedule of treatments with a particular individual. Therefore, in terms of retrieval, these three parts of the client profile will be stored separately and will require a "key code" to reconcile them.

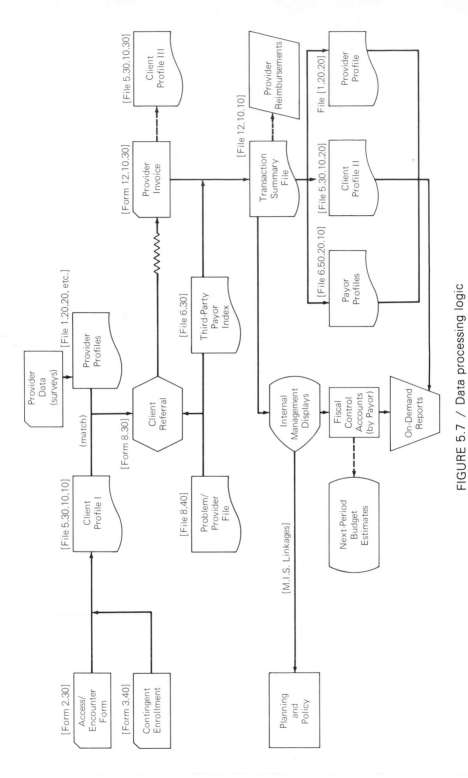

FIGURE 5.7 / Data processing logic

At any rate, the qualification screeners (perhaps using data supplied by an outside access agent) develop the first part of the client profile, which is then used to make a determination about the client's eligibility. Assuming that the client is indeed eligible, the information system might encounter him again when a treatment referral is to be made. The information system, in making a referral, is equipped with information about all regional (or cooperating) providers, and would thus provide this information —along with the client data—to the referral algorithm (as discussed in the final section of Chapter 4). The client is then given a referral (form 8.30), and the information system then awaits the return on an invoice from the provider. This *invoice* tells what treatments have been performed (and also requests reimbursement, and may also indicate the need for a further referral). The information on the invoice (form 12.10.30) is then transferred to Part III of the client's essential medical record, if appropriate, and then becomes evidence of a *transaction,* per se. The transaction logic, as was explained earlier in this chapter, sees that the provider is properly reimbursed for his services and then acts to debit the appropriate payor account. The parameters of the transaction are also used to update the Part II of the client's profile, showing what treatments he has received, etc., and also the rate at which he accepts (and effects) the referrals recommended for him. Other data goes to update the provider's profile, allowing us to maintain an up-to-date record of the amount of HCDS business that provider is doing (so we can avoid a concentration of referrals). Finally, the transaction data is kept available for entry into reports required by payors and also to support internal management and planning-forecasting decisions to be made by the system manager. Finally, associated with the referral function, there is the file mentioned in Chapter 4 that establishes what providers are qualified for what treatment-diagnosis situations (e.g., the coded file of specializations), and also an up-to-date file that we can scan to see if there are any third-party payors that might pay for this particular treatment. This latter provision is in addition to the notification of standard third-party payors which are always indicated on Part I of the client profile itself. The use of the third-party payor file (file 6.30) is to see if the *treatment* rather than the client, per se, is one that some other agency might support, and also allows us to keep a constant check on the extent to which already enrolled clients may become available for third-party payor coverage as time passes.

This very broad overview of the data processing functions of the HCDS must be elaborated and carried to more detail (at least for some of the information processing mechanisms). But before doing this, the reader should get an idea of the structure of the forms and files that are operated on by the information subsystem. This is the function of the next section.

DOCUMENTS AND FILE STRUCTURES

A data processing system produces information. Information, per se, is always the product of some components of the (raw) data base being manipulated by some processing algorithm. The dynamic logic of the HCDS information subsystem thus is reflected in the nature of the processing algorithms. Before these can make sense, we must take a look at what it is the algorithms manipulate. In short, we must know something about the HCDS data base, which means documents (qua records) and file structures. After all, every substantive function of the HCDS or a social service system implies an informational or data predicate. So, as we now turn to the anatomy of the HCDS data base, we also have an opportunity to refine some of the arguments about access, referral and other functions from earlier sections. We shall begin these discussions by taking a look at the forms (documents) used by the system, and shall then discuss the files into which these forms are evolved.

Forms and Their Implications

Every transaction that a social service system effects—and every client contact—gives rise to what is called a *record*. When records are collected together (in files of some sort) or otherwise ordered, codified and summarized, there is the basis for a history of the system during some interval. To the system analyst, then, a list of the forms associated with a system tells a great deal about the substance and dynamics of that system.

In the pages that follow, then, we shall take a look at what forms the HCDS system uses in what functions (with the relationship clear from the fact that we have indicated form numbers on all of the referent logic charts we have reproduced), and also show how those forms were structured to facilitate data manipulation processes. Just for reference purposes, the schedule of the forms to be discussed is given below:

Form Number (Task No.)	Title
2.30	Encounter/Access
5.20.10	Personal and Qualification Data
3.40	Contingency Enrollment
5.20.20	Administrative/Referral/Treatment
5.20.30	Essential Medical History
7.80.40.10	Medical Screening/Referral/Invoice
8.30	Referral Notice/Voucher
11.10	Request for Client Records
12.10.30	Invoice Form
14.50	Suggestion Form
2.40	Agency Referral
4.60	Accounts Receivable Billing
4.80.10	Exception Notification
7.70.40.20	ID Card/Contract Form
12.50	Notice of Nonresponsibility
14.10	Notice of Inability to Make a Requested Referral
15.0	Provider Reimbursement Statement
16.0	Notice of Suspension of Client

In analyzing the structure and role of these various forms, it is useful to separate input from output documents.

INPUT FORMS

Input forms are those documents (or records), generated either within the HCDS or external to it, that *cause the system to initiate some action.* We may briefly now see just what the input forms look like and what responses they dictate.

FORM 2.30—ENCOUNTER/ACCESS FORM: This form is distributed to all those agencies or individuals which might serve as access points for system clients. These would involve, among others:

Agencies of the federal, state or local government

Private physicians, hospitals or government or private clinics.

Specially qualified individuals such as priests or ministers, parole or police officers, members of the judiciary, field workers for private social agencies (e.g., YMCA, CYO, United Way, American Cancer Society)

The information to be completed on this form by the access agent should bear directly as possible on the two key decisions which must be made with respect to any client:

Whether or not, given existing threshhold conditions, he is eligible for enrollment and medical support (and to what extent, if there are any restraints).

Into which of the various coverage modalities the client should be placed (block, fee-for-service, HMO, insurance) if indeed he should qualify for enrollment.

The data, then, asked for on this form should be that which would lead to the association of the particular client with one or another of the socioeconomic and medical categories defined in Chapter 4.

It is expected that most access agents, and particularly local service or field agents, will be able to supply reasonably well validated information about the key socioeconomic properties of the individual:*

Residency status

Economic profile (including, if possible, data on income and expenses, etc.)

Social profile (the agent's assessment of the proposed client's social and behavioral properties)

Cognitive class (an objective assessment of the client's cognitive capabilities and degree of self-responsibility, etc.)

Mobility (the extent to which the client has his own transportation, etc.)

In some cases, the information used to complete the access/encounter form might best be completed in the client's absence. Anyway, this portion of the form might best be taken directly from the definitions given in Table 4.2 and transferred onto Form 2.30 as follows:

*Data about the third-party coverage the client has would also be useful.

Socioeconomic Summary

   (Check one of the boxes for each of the socioeconomic dimensions)
   1. Residency: ____Transient _X_ Semiperm.
                    ____Long-Term ____Fixed
   2. Economic: ____Unemployable _X_ Structurally Unem-
                    ployed ____Working Poor ____Retired/Disa-
                    bled

When these access/encounter forms are sent to various access agents, an instruction booklet should accompany them so the agent may have recourse to the formal definitions of each of the above categories given in Chapter 4.

Where the access agent is medically qualified (public health nurse, physician, or emergency room intern*, it may be possible to get an adequate assessment of the medical classification or category to which the client would belong were he enrolled in the system—*episodic, temporal, secular, chronic.* Again, the access information booklet which would be sent to all parties receiving access/encounter forms would indicate the formal definition or criteria for each of these four medical categories, and the access agent would be asked to indicate that particular category which most nearly approximates his assessment of the client's medical condition.

It should be noted that socioeconomic profiles and medical classifications done by access agents would be given more or less weight depending on the particular qualifications of the agents. Thus, for example, a social worker's socioeconomic assessments would generally carry a high presumption of accuracy, as would a physician's medical assessments, with correspondingly less weight given items completed by nonspecialists.

For practical purposes, those assessments given high weights on the basis of the access agent's presumed expertise would need not be reperformed by HCDS personnel. On the other hand, items completed by nonspecialists (or merely left blank because the agent had not the requisite information) should be completed by the HCDS officer under *Form 5.20.10,* discussed below.

FORM 5.20.10—PERSONAL AND QUALIFICATION DATA: This form would contain essentially the same informational items as Form 2.30, but would contain additional informational items for *internal* use.

   Validated economic criteria on client pertinent to determining eligibility

*This form would be filled out for emergency treatments under the HCDS Contingency Enrollment Program (see Form 3.40)

Information on any third-party payors who may have an interest in this client

A place for the prospective client's signature signifying the truth of the items contained on the completed form pertinent to eligibility, perhaps with a penalty notification for false assertions

A tear-off portion which may be transmitted to the original access agent indicating HCDS's disposition of the case.

This document would have legal implications and be the basic operational form for enrollment determination, containing data such as the client's address, telephone number, social security number, and whatever socioeconomic or medical data was collected by an access agent, in addition to the items mentioned above.

Thus, in some cases, *Form 5.20.10* would simply be a matter of copying from the *Form 2.30* filled out by an external access agent, thus minimizing the amount of administrative energy to be expended by HCDS, per se. Thus, it is important that the *Form 2.30* be distributed as widely as possible, and instructions developed such that as much of the data collection as possible be performed by those in the best position to do so with a minimum of cost or time or error (e.g., those closest to the client and his community).

Note: This form becomes the direct input to File 5.30.10.10 (see next section).

FORM 3.40—CONTINGENCY ENROLLMENT: This form helps implement one of the more interesting and unusual aspects of the broker system. The contingency enrollment option reflects a standing commitment to all medical providers that any validated emergency services performed on a casualty will be indemnified by HCDS, providing that the casualty has neither sufficient personal resources nor adequate third-party coverage. Under this provision, which should be considered a primary public relations element, no citizen (indigent or otherwise) need fear being refused the best and closest medical attention in the event of an emergency.* Thus, this form will be made available primarily to hospital emergency service units. The patient, when able, will be asked to sign it. In substance, the form simply indicates that the patient agrees to repay to HCDS any expenses incurred on his behalf (or to arrange such payment from any third-party payor who may cover him) should he subsequently be found to be ineligible (because of any criterion) for enrollment as a regular participant in the program. This Contingency Enrollment Form would be forwarded to HCDS and would contain essentially the same qualification data as *Form 2.30*. Where services have already been performed and the patient has been released, a *Form*

*See, again, the second section of Chapter 4.

12.10.30 would accompany the Contingency Enrollment Form. It will be the responsibility of the hospital, in their normal course of business, to elicit information from the patient on any personal resources or third-party payor coverage. A formal definition of medical emergency should be supplied all providers receiving this form.

FORM 5.20.20—ADMINISTRATIVE/REFERRAL/TREATMENT:* This form, along with *Form 5.20.10* and *Form 5.20.30,* constitute the *client profile.* This form is used to record referrals which have been made on behalf of the client, the disposition of those referrals (e.g., whether they resulted in treatment), the nature, dates and cost of treatments, and the provider performing them (and, where appropriate, the particular payor debited). For a manual system, there would have to be a "protected" code which would be used to relate a patient's form *5.20.10, 5.20.20,* and *5.20.30,* and this might generally mean that the three documents comprising the patient or client profile would be kept in separate files. (For the computer-based system, see the way these files are treated under the paragraph entilted *Client Profile,* in the next section on Files.)

FORM 5.20.30—ESSENTIAL MEDICAL HISTORY:* This form would be used to record those diagnoses, treatments, and other data (e.g., blood type, allergies) which would constitute the essential medical history of a client. This form would, in effect, simply be used to abstract pertinent items from *Form 5.20.20,* or data taken from a medical screening (see below) or information made available from any other sources. Under certain circumstances, where requested by providers, copies of the essential medical record could be forwarded so complete histories need not be constantly retaken. The disposition, protection and substance of such forms should be determined in consultation with both medical and legal authorities.

FORM 7.80.40.10—MEDICAL SCREENING/REFERRAL/INVOICE: This form would be the authority (e.g., voucher) for the performance of a medical screening on a client. These screenings would generally be undertaken under one of two circumstances.

> When there is sufficient doubt about the medical classification into which a client should be placed, pertinent to the decision about the most appropriate coverage modality

> When it is deemed by some responsible system authority economical to screen an individual, in the hopes of being able to apply preemptive care which will result in a net savings to the system (or in a significant net benefit to the client himself)

**Note:* These forms would not be needed were the patient profile files computerized.

This form should have space for the examiner to record the results and suggest treatments he may consider necessary (which would then be translated into specific client referrals). The return of this form, completed, would constitute an invoice.

FORM 8.30—REFERRAL NOTICE/VOUCHER:* This form is issued for a specific referral or set of referral alternatives to be made for a client. For clients under the block coverage modality, the form will authorize a particular provider to supply treatment for a specific condition, perhaps citing a ceiling payment level, for example, for episodic conditions. In other cases, the form may notify a client who has requested assistance in selecting a provider of the particular provider(s) deemed most appropriate (these being considered most cost-effective from the standpoint of HCDS). Where appropriate, the form may specify a certain appointment time and date. Where the request for a referral is made on behalf of some client by an authorized agent (e.g., a public health nurse, a welfare case worker, a private physician), the completed referral form will be sent to that agent for transmittal to the individual. As a public relations aspect, the referral service may be made generally available to all residents, agencies, medical providers, etc., as individuals (and particularly new residents) might have difficulty in making provider-oriented decisions. Making this service generally available may help ease instances where erroneous decisions are made and might assist in preserving the integrity of the medical community in the region. (For a discussion of nonmedical referrals, see the section describing the use of *Form 2.40.*) Where this form serves voucher purposes, a tear-off portion should be returned by the authorized provider along with his invoice for treatment *(Form 12.10.30).* A note of the referral, in all cases, should be added to the client's *Form 5.20.20.* It need hardly be mentioned that the extent to which clients make use of the referral option will have a great bearing on the overall cost-effectiveness posture which the system is able to maintain. Therefore, as part of the enrollment agreement, clients should be notified and asked to acknowledge by signature the fact that treatment by nonauthorized providers will nullify the system's obligations for payment; therefore, when in need of service, the client would do well to ask for a referral or check to see that his physician or provider of choice is HCDS authorized. In this latter case, the form would be returned to the client with an indication that the referral is acceptable. Where the provider of choice is unauthorized, the system should generate a referral or list of system-preferred alternative providers. (See notes on *Form 7.70.40.20.*)

*Note: It may be desirable to distribute copies of this form to access agents and providers in the field, along with *Form 2.30* and invoice forms, etc.

FORM 11.10—REQUEST FOR CLIENT RECORDS: In some cases, providers may request the essential medical history of an enrolled client *(Form 5.20.30).* This form initiates that request. A check should be made to validate the authenticity of the request and the fact that the provider is an HCDS authorized one.

FORM 12.10.30—INVOICE FORM: Whenever a provider performs treatments or services for an enrolled client, he will prepare this form which will serve to apprise us of costs of service, the diagnosis (or problem) treated, the schedule of treatments performed and whether or not further referrals should be made for this client. Diagnosis and treatment designations should be in the appropriate code(s) (see Task 6.60).* Costs should be scheduled by treatment. The provider may also indicate information which should become part of the client's essential medical history. Treatment, diagnosis and cost data should be transferred to the client's *Form 5.20.20,* along with dates and provider identification. The form then goes into the Reimbursement Subsystem (see Task Section 12.00). The information on the invoice will also be entered in the appropriate place on the Provider File and Payor File (see these file descriptions in a later section).

FORM 14.50—SUGGESTION FORM: This form will simply serve to allow clients, system components, or any other user of system services to bring to the attention of the system manager dysfunctions in the system; they will facilitate, as it were, an ombudsman subsystem.

OUTPUT FORMS

These are forms generated for external use (via reporting requirements) or which are kept as part of the internal control function. In either case, they are results of system actions, rather than initiators of action.

FORM 2.40—AGENCY REFERRAL: Wherever a client (or prospective client) or individual is deemed by system authorities to be a candidate for nonmedical treatment or services, the individual will be referred to the agency most appropriate (as determined with reference to the Agency Information Index, *File 2.20.20).* Where the condition for which the client is referred is a reportable one, under the legal definition of that term, a copy of the referral will be kept in the *Action File (2.60.10),* and this file will be periodically reviewed to see that referrals were actually exercised by the client. In all cases, a copy of the referral

*Recall that all system design and implementation "tasks" were set out in Chapter 2.

will go to the agency in question as well as to the client, along with whatever information the system officer can provide to justify the referral.

FORM 4.60—ACCOUNTS RECEIVABLE BILLING: This form, specifying an amount owing (or perhaps an installment payment on an outstanding account for a specific period) will be sent under two conditions:

> Where the client received emergency treatment under the contingency enrollment provision but was subsequently deemed ineligible for payments made by HCDS on his behalf

> Where the client is a copayor for medical services

In either case, forms will be copied into the *Accounts Receivable File (4.40.10)*, and periodically examined for compliance.

FORM 4.80.10—EXCEPTION NOTIFICATION: This form will be sent to prospective clients who failed to meet the criteria for eligibility and enrollment. The detail of explanation will be determined by the system manager.

FORM 7.70.40.20—IDENTIFICATION CARD/CONTRACT*: A card uniquely identifying the client will be sent to all individuals accepted for eligibility by the system, except those entered into the block coverage modality. This card will specify the type of provider who will be indemnified with respect to the specific individual (e.g., HMO, all physicians authorized by the insurance carrier particular to the client), and will set forth any constraints (e.g., a client may be covered for a work-related condition by some third-party payor; the ID card would therefore notify providers that HCDS will not pay for such services). In addition, cards will carry an expiration date, which will vary with the coverage modality to which the individual has been assigned (e.g., fee-for-service clients may have cards good for six months). In cases where the client is covered by an HMO or insurance carrier, he may be provided only with the card from that coverer, such that ID cards will primarily be issued only to fee-for-service clients, or to those for whom HMO or insurance coverage is inadequate. In all cases, a client will be required to sign a contract setting forth his understanding of limitations and system liabilities; e.g., as explained under the paragraph dealing with *Form 8.30*, clients should be asked to notorize the condition that the system will not pay for services performed by an unauthorized provider. For clients receiving the ID cards, the signing of the card will be understood to indicate comprehension and acceptance of the contract conditions. In this respect, it is useful to send this form to all but

**Note:* This card should also indicate copayor status, if appropriate.

block clients, even if the client will have another card from an HMO or insurance carrier.

FORM 12.50—NOTICE OF NONRESPONSIBILITY: This form will be sent upon receipt of invoices *(Forms 12.10.30)* from providers:

Who are not HCDS authorized

Who have provided services to an individual who is not a system client. In this latter case, a blank Form 2.30 will accompany this form, so the provider may fill it out and propose his patient for eligibility. If the patient is eventually accepted, the invoice may be reprepared and reforwarded.

FORM 14.0—NOTICE OF INABILITY TO MAKE A REQUESTED REFER-RAL: This form will be sent to a requester for a referral when the system is incapable of meeting that request because of the lack of a suitable provider, etc. A copy will be stored in *File 20.0;* items in this file will be periodically reviewed to indicate where gaps are in provider coverage and distribution.

FORM 15.0—PROVIDER REIMBURSEMENT STATEMENT: This form simply serves to collect the statement of payments due a provider during a specific reimbursement cycle. It should list transaction number, diagnosis code, date and patient name, along with the payment for that client/date/diagnosis combination. Also, the form should be able to tell the provider of any adjustments in originally invoiced amounts due to P.S.R.O. or other audit functions. It is on the basis of this statement that payment warrants will be drawn, forwarded along with this statement to the provider. It is suggested that this form be a fold-over, on which a computer-generated address label may be fixed, with a pocket for the warrant. On a statistical sampling basis, a test should be made between incoming invoices and back copies of these statements to try to minimize the possibility of providers charging more than once for the same services. A repeat of a client/date/diagnosis combination would indicate such an attempt.

FORM 16.0—NOTICE OF SUSPENSION OF CLIENT: A client who has abused the system will be sent this notice, and his HCDS status code on his profile will be changed to a code "5."

Files and Their Manipulation

As was previously suggested, all documents or forms create records. Some of these records are then organized (or stored) in files. The files used by the HCDS are listed below (note that the file numbers correspond, as did the form numbers, to components of the master logic charts and the task numbers given in the task schedule section in Chapter 2.):

File #	Designation
1.20.20	Provider File I (Physicians)
1.20.30	Provider File II (Hospitals)
1.20.40	Provider File III (Clinics)
1.20.50.10	Provider File IV (Nursing Homes)
1.20.50.20	Provider File V (Homes for Aged)
2.20.20	Agency Information Index
2.60.10	Action File
4.20	Contingency Enrollment
4.40.10	Accounts Receivable
4.80	Excepted Clients
5.30.10.10	Client Profile I (Personal/Qualification)
5.30.10.20	Client Profile II (Administration)
5.30.10.30	Client Profile III (Essential Medical History)
6.30	Third-Party Payor Index
6.50.20.10	Payor File I (Medicare)
6.50.20.20	Payor File II (Medicaid)
6.50.20.30	Payor File III (Welfare)
6.50.20.40	Payor File IV (Categorical)
7.80.40.20	Screening Suspense File
8.40	Problem/Provider Category Index
12.10	Transaction File
12.80.10	Holding File
12.10.10	Transaction Summary
8.30	Archive File for Referrals
14.0	Wait-List for Prospective Clients
20.0	Holding File for Form 14.0

The HCDS forms were separated into input and output types. For files, a more useful distinction is between *manual* files (those which might reasonably be kept on something other than on-line or immediate computer storage) and *computer-based* files, these latter expectedly being accessed and manipulated with some frequency. If no computer facilities are available to a program—because of resource limitations or the small client and provider populations, etc.—then all files may be generated and manipulated without benefit of a computer. In the early stages of a program, before the enrollment has reached significant levels, a manual operation is recommended. Then, as client enrollment builds, an audit of actual transactions can suggest the frequency with which various files are used, and therefore may disclose the best schedule for computerizing the files (which implies the best method of allocating scarce administrative overhead resources). At any rate, we can begin with a brief description of the manual files.

MANUAL FILES

The following files are simply repositories for forms or relatively infrequently accessed and/or altered (updated) information. It is anticipated that these would be computerized only were a significant influx of resources dedicated to such a task made available to HCDS.

FILE 2.20.20—AGENCY INFORMATION INDEX: This file would actually be a simple list of all the external agencies with which the HCDS should have a reciprocal agreement to exchange information and/or clients. This index might simply be an amplified and perhaps composited adaption of social service indices already existing in local communities, and should be updated periodically to reflect new agencies or programs moving into the locality (or to erase programs that have been deactivated). As an adjunct, this should be the referent for the development of the type of cross-correlations (casualty clusters) discussed in Chapter 4, showing the frequency of shared clients among the various community agencies and programs. Again, any single program may initiate the development of such an index, but might do well to invite participation by other units in an effort to share developmental costs.

FILE 2.60.10—ACTION FILE: This is used to retain *Forms 2.40* for reportable conditions, subject to periodic survey to determine client compliance or responsiveness. Items stored here may be pulled and discarded (or sent to an archive file) when the disposition of the case by the agency referred is received by HCDS (the tear-off portion of *Form 2.40*).

FILE 4.20—CONTINGENCY ENROLLMENT: This houses contingency enrollment forms *(Form 3.40)*. These forms will not be acted upon until an invoice has been received from an emergency provider pertinent to the individual in question. If no invoice is received, we may assume that no financial obligations will fall to HCDS. Items may be purged after one year and should be cross-referenced therefore by name and date.

FILE 4.40.10—ACCOUNTS RECEIVABLE: Held here is *Form 4.60* (A/R Billings) for both copayors and contingency enrollment clients found ineligible but for whom payments have been made or invoices received (as the contingency enrollment option indemnifies providers against nonpayment for emergency services). These should be handled according to normal A/R logic, with several stages of collection tactics.

FILE 4.80—EXCEPTED CLIENTS: Duplicates of *Form 4.80.10* sent to clients who have been declared ineligible are in this primarily archival file.

FILE 6.30—THIRD-PARTY PAYOR INDEX: This file houses the information pertinent to all third-party payors who might have an obligation (legally or by precedent) for certain categories of clients. Before a referral is made for a client, every effort should be made to exhaust all third-party payors who might obviate HCDS payment, and clients will be screened for any third-party payor opportunities at their entry into the system (see Tasks 6.20 and 9.10.30). This file should be organized by client category (e.g., aged American Indians, juvenile delinquents), and for each third-party payor the index should contain current data on limits of coverage, substance of coverage, categories of coverage, eligibility and application procedures, any constraints.

System referral personnel and management would be urged to familiarize themselves with the opportunities here, and the referral officer should be required to note that he has explored third-party payor sources before initiating a client referral for an eligible client.

FILE 7.80.40.20—SCREENING SUSPENSE FILE: This file would be used for copies of *Form 7.80.40.10* until completed forms have been returned by the screener. The *Form 2.30* and *Form 5.20.10* associated with clients to be screened would also be held here until a coverage modality is determined for the client and his coverage is initiated. At that time, data on the *Form 7.80.40.10* would be transferred to the appropriate places on the client's *Form 5.20.20* and *5.20.30*. Thus, until a client for whom a screening was suggested actually completes the process, he is not eligible for HCDS coverage.

FILE 8.40—PROBLEM/PROVIDER CATEGORY INDEX: This file is a simple index which will identify all the various categories of problems (e.g., diagnoses, medical conditions) for which HCDS is legislatively responsible. Associated with each of these categories of coverage will be the category of provider most appropriate (e.g., for muscular dystrophy, refer the client to a neurologist; for heart murmur in an infant, refer client to a pediatric cardiologist). This index will be used in the completion of *Form 8.30* and will enable nonmedically qualified individuals (or, eventually, a computer) to make appropriate referrals.

FILE 12.10—TRANSACTION FILE: This file is the basic disbursement file and houses the invoices *(Form 12.10.30)* returned for payment during some period—two weeks would probably make providers happier than a one-month period, etc. Transactions (invoices/disbursement/recordings on provider, client and payor files) will be cleared and the transaction forms may then be retired to an archive according to the logic developed under Task Area No. 12 (diagrammed in the project logic charts).

FILE 12.80.10—HOLDING FILE: This file contains those invoices on which a P.S.R.O. audit is outstanding or which are to be audited prior to payment. Upon receipt of the audit, the invoiced amount will either be paid or adjusted and the invoice moved into place in *File 12.10* for retirement at the next payment period.

FILE 8.30—ARCHIVE FILE FOR REFERRALS: This file simply houses referrals made on behalf of system clients, and may periodically be reviewed to indicate the number of referrals made by category, etc. By examining the nature of the referrals requested, the system manager will be able to get an idea of where there are lacks of information in the community about medical services available, etc.

FILE 14.0—WAIT-LIST FOR PROSPECTIVE CLIENTS: Here are held the qualification forms *(Form 5.20.10)* for clients who are expected to become eligible for enrollment at some future time (e.g., when resources are increased or as other clients retire from eligibility). The file should be prioritized according to HCDS status codes (1 = priority, 2 = preferred, 3 = marginal), which appear on each clients profile.

FILE 20.0—HOLDING FILE FOR FORM 14.0: This file serves to hold all *Form 14.0s*, which indicates that a request for a referral has not been able to be filled (see discussion of *Form 14.0*).

COMPUTER-BASED FILES

Following are descriptions and details pertinent to those files which, at implementation of the HCDS, are expected to be housed on a computer system. Included, primarily, is formatting and manipulation data.

The provider Data Base

FILE 1.20.20—PROVIDER FILE I (PHYSICIANS): This is the basic reference file for referrals for ambulatory care, for fee-for-service and insurance clients. It will contain the data items listed under the format below. The BME license no.* should be the unique identification code for each physician included. Physician listing should be by category (e.g., those developed for *File 8.40*), and sorted by BME no. within categories to ease tape-based data processing.† (Other sort information is given in the format following.) Whereas data fields 1 through 8 are self-explanatory, the subsequent fields should be explained in a bit more detail (refer to format below).

Field 9: In this field, we will place a four-digit HCDS Audit Index No., developed according to the logic set out in Chapter 4. It is recommended that, at the initiation of the system, each physician be given an index number valuation of 10.0, with subtractions for causes mentioned in the above sections. The amount detracted will be determined by the system manager, but might be scheduled beforehand for general cases of overtreatment or overcharging, etc.

Field 10: This field will record the aggregate number of transactions associated with this provider up to the current accounting period.

Field 11: This field will record the aggregate payments made to this provider up to the current accounting period.

Field 12: This field will record the number of transactions in which there have been incidents resulting in a reduction of the provider's index number value.

FILE 1.20.20 (FIELD FORMAT)

Field No.	Size	Data Item
1	5 (99999)	BME License No.
2	15X	Name
3	20X	Street Address
4	20X	City, State, Zip Code/County Code

*The BME number would be a designation assigned by the State Bureau of Medical Examiners or some other licensing authority.
†The same essential formatting operations would be performed were, as is likely in many cases, the files to be kept on some sort of direct access medium (e.g., disk, drum).

5	5 (999V99)	Census Tract/CCD (99XXX)
6	2X	Primary Specialty
7	2X	Secondary Specialty
8	2X	Group Affiliations
9	4 (9999)	HCDS Audit Index
10	3 (999)	# Transactions (Aggregate)
11	8 (99999.99)	Aggregate Payments
12	2 (99)	# Incidents
(13-n)	7 (9999999)	Transaction Codes

Provider File I (Physicians)

Major Sort by Specialty/Category
Minor Sort by Census Tract (CCD)
Minor Sort by HCDS Audit Index
Inverse Sort by # Transactions (to distribute referrals)

These fields are all optional, and depend on whether or not annual or semiannual composite reports or summaries are produced by providers. If they are, then the fields would contain the values of the previous period. A separate report could then be produced for the system manager which would enable him to compare the period-by-period performance of providers (in an effort to spot deteriorations in service or excessive reliance on brokered clients, etc.).

Fields 13-m: These will be members of a variable length record in which the code number for each of the transactions in which this provider has been involved will be recorded. This is necessary for audit purposes (see the discussion of *File 12.10.10* under Transaction Data Base). The transaction schedule should be updated with each disbursement cycle.

FILE 1.20.30—PROVIDER FILE II (HOSPITALS AND OTHER IN-PATIENT FACILITIES): Here we would establish a file containing pertinent data on all hospital and in-patient facilities, with data obtained on geographic location, nature of special facilities (e.g., special diagnostic and treatment capabilities), and patient restrictions (e.g., pediatric, geriatric) or conditional restrictions (e.g., cardiac, eye and ear conditions). As with physicians, we would want to include here all the information necessary for making rationalized referrals. Therefore, there would be an audit (cost-effectiveness) index, a count of prior referrals to keep in-patient referrals distributed among different facilities, an indication of the payment modality demanded (a straight per diem, a diagnosis-related reimbursement, etc.), as well as an indication of any block provisions that might have been prepurchased in that institution. The format used would also have to keep accounts on the cumulative billings, disputed reimbursement requests, patient comments about service, etc.

FILES 1.20.40, 1.20.50.10, 1.20.50.20—CLINICS, NURSING HOMES, HOMES FOR THE AGED: Essentially the same information demanded for rationalized referrals to hospitals would be required here. The formats of these files, therefore, should be similar to those used for hospitals, etc. Free clinics, however, would be listed under the third-party payor index *(File 6.30)*, and other clinics would be sorted primarily by specialty (or class of patient), followed by the normal minor sortings on geographic area (as given by the census tract), HCDS audit index and an inverse sort on number of transactions to spread referrals among otherwise equal provider facilities. Essentially, the same logic would hold for the files on homes for the aged and nursing homes.

　　Note: The five files constituting the aggregate Provider File should each be placed on separate tapes, but thought of as constituting a data base, per se. Should random access facilities become available, the basic segmentation of the data base into the five basic provider categories should be maintained, but internal sorting logic could be relaxed or modified.

THE PATIENT PROFILE DATA BASE

The patient profile data base will maintain records on all HCDS clients. For each client, there will be three logical records which, for purposes of security, will be separately located (e.g., on different tape files), with the identification numbers protected by a functional encyphering (the identification code for a client's administrative record and essential medical history is a "secured" function of the code identifying him on the personal and qualification data file—*File 5.30.10.10*). Thus, locating a client's administrative and referral record in *File 5.30.10.20* would not enable a correlation with a particular individual unless the searcher knew the functional code in use. Each of the three files will now be discussed in sequence.

FILE 5.30.10.10—PERSONAL AND QUALIFICATION DATA: The format and data comprising this file is given below. This file will contain records of all individual's with data being taken primarily from *Forms 2.30* and *5.10.10*, with the information capable of being updated at any time on an edit run. The data items are self-explanatory, with the exception of Field 22 which simply summarizes the client's status within the HCDS system for use whenever a cutback in enrollments is dictated by fiscal conditions, e.g., those with a priority rating would be last to go. Suspended clients would be those who, perhaps for failure to pay copayor obligations, are currently ineligible, etc.

FILE 5.30.10.10 (TAPE FORMAT)

Field No.	Size	Data Item
1	N(X)	Indentification Code (XX Payor; S.S. No.)
2	15X	Name
3	20X	Street Address
4	20X	City, State, Zip Code/County Code
5	5 (999V99)	Census Tract/CCD (99XXX)
6	10 (9)	Telephone Number
7	3 (9)	Age
8	X	Race
9	X	Sex
10	X	Income Class
11	X	Marital Status
12	1 (9)	Residency Class
13	1 (9)	Social Profile
14	1 (9)	Economic Profile
15	1 (9)	Cognitive Class
16	1 (9)	Mobility Class
17	1 (9)	Medical Classification
18	2 (9)	Family Size
19	1 (9)	Coverage Modality
20	8 (99/99/99)	Expiration of Eligibility
21	5 (99.99)	Copayor % (ϕ if n/a)
22	1 (9)	HCDS Status (1 = priority; 2 = permanent, 3 = marginal, 4 = excepted, 5 = suspended, 6 = contingency)
23	M (2X)	Third-Party Coverage (Codes from File 6.30)
24	N (NX)	Coverage Restrictions (Codes from File 8.40)

FILE 5.30.10.20—ADMINISTRATIVE AND REFERRAL DATA: This file will contain the second segment of the client profile, that giving a codified summary of his history or utilization in the system (the client "track"). These records will be uniquely identified by a modified identification code (a function of Field #1 in *File 5.30.10.10*). The records should be sorted on the Identification Field and will have a simple format like the following:

FILE 5.30.10.20 (TAPE FORMAT)

Field No.	Size	Data Item
1	9 (X99999999)	Modified Identification Code
2	1	HCDS Status
3	8 (99/99/99)	Eligibility Expiration Date
4	5 (99.99)	CoPayor % (if none = ¢)
5	m (2X,NX)	Diagnosis Code/Third-Party Payors*
6	3 (000)	# Transactions for Client
7	9 (99999.99)	Aggregate Expenditures for Client
(8 − m)	nX	Problem/Diagnosis Code (File 8.40)
	7 (9999999)	Associated Transaction Code

The data here will permit the system manager (or, eventually, a computer) to track the patient's history in the system and to validate any invoices as to the eligibility of the client for the particular treatments given. Thus, invoice validation and entry of a transaction can be done in one step. The diagnoses/problem codes are broken out so that the frequency and number of conditions demanding treatment may be examined without having to go to the transaction archive. Note that if HCDS status, eligiblity expiration date or coverage restrictions are not in order with respect to a particular invoice, the provider will be sent Form 12.50 and the transaction code flagged for that diagnosis.

FILE 5.30.10.30—ESSENTIAL MEDICAL HISTORY: This file simply records data taken from *Form 5.20.30* or, in a computerized system, will be updated automatically from completion of a medical screening or when a returning treatment invoice is flagged as having data pertinent to the EMR. Again, as with the previous file, records here will be protected by having an identification number which requires a two-step functional resolution, that is, the identification code for records in this file will be some function of the Identification Code for *File 5.-30.10.20* which in turn, is a function of the Identification Code in *File 5.30.10.10*. Records in this file should have a leader which notes destinations (providers) to which copies of a client's EMR were sent. The format for this file must be developed in conjunction with a medical expert who can give information on what items should constitute an EMR, aside from such obvious factors as blood type and allergies to drugs, etc.

The payor data base will be comprised of five files:

Note: The Diagnosis Code/Third-Party Payor pairs indicate coverage for which HCDS need not pay, and should be used in the referral and transaction processes to reduce system costs.

1. File 6.50.20.10	I—Medicare
2. File 6.50.20.20	II—Medicaid
3. File 6.50.20.30	III—Welfare
4. File 6.50.20.40	IV—Categorical
5. File 6.50.20.50	V—General

The files all exist primarily for audit purposes and will be used mainly to record the transaction code numbers for those transactions which ultimated in a debit to the particular payor's account (e.g., a reduction in the residual budget through a disbursement associated with an expenditure for a client associated with a particular payor). Thus, to all intents and purposes, each of these files will constitute a variable length logical record which is formatted as follows for *Files 6.50.20.10–6.-50.20.40:*

GENERAL PAYOR FILE FORMAT

Field No.	Size	Data Item
1	10 (9)	Budget Amount (Aggregate)
2	5 (9)	Capitation Number (if appropriate)
3	2 (9)	No. Two-Week Periods Budgeted
4	4 (9)	No. Transactions (aggregate)
5	7 (9999.99)	Cost per Transaction (mean)
(6–31)	5 (99.99)	% Budget Expended/2-week period (cumulative)
(32–57)	5 (99.99)	% Capitation Exhausted/2-week period (cumulative)
(58–n)	7 (9999999)	Transaction Code
	8 (99999.99)	Associated Expenditure

The determination as to which of the various payor files is to debited for a particular transaction is determinable for most clients by the first digit of the identification code $x99999999$, from the format for *File 5.30.10.10.* The "x" should be coded to indicate to which payor category the client belongs. Thus, when the invoice for treatment returns for payment, the Identification Number will be used not only to locate a particular client's Administrative and Referral Data record (in *File 5.30.10.20*), but also to indicate the particular payor file which should be charged with that particular transaction. For clients who are not clearly the responsibility of a particular payor, or where certain payors cover only certain conditions for the client, then a manual override option must be exercised before a payor account is debited. As for *File 6.50.20.50,* this would simply be a standard cost accounting file into which administrative and overhead accounts would be placed.

THE TRANSACTION DATA BASE

The transaction data base is actually a composite of three different files, two of which have already been defined: *File 12.10* (the transaction file) and *File 12.80.10* (the holding file). These two files are actually operational files, that are operated on every two weeks (the reimbursement cycle), and then purged. The file we shall define here, however, is an aged archival file organized by fiscal year.

FILE 12.10.10—TRANSACTION SUMMARY: When a transaction form is returned from *File 12.10,* it is placed in this file. The current segment of this file will maintain all invoices (transaction *Form 12.10.30*) for the current fiscal year, as these will be the primary audit trail for the HCDS. Note that in the provider, client and payor files just discussed, the primary unit of information is the transaction code, a seven-digit number. Each invoice *(Form 12.10.30)* is given a unique transaction code; in addition to the diagnosis, treatment, cost and further referral information supplied by the provider, there will be the client number (which enables us to connect with the Client Profile File II *(5.30.10.20)* to record the transaction number there, while the "x" code enables us to identify the payor file to which the transaction code should be added. The provider will also have a unique identification code (Field #1 on *File 1.20.20*), thus the transaction code is added to that file.

At the end of the fiscal year, or at any time that a summary accounting is required, the transaction codes in the payor, provider and client files enable us to precisely reconstruct the history of the system as a whole during the relevant period, or to summarize the activities of any payor, provider, or client. The use of this semiarchival file also enables us to conserve space on the operational files by not recording the diagnosis/treatment/cost/date, etc., more than once.

The only major data processing activity, then, would be the construction of major summaries from the transaction summary file for each major accounting period. The format of this file is given below, the *Form 12.10.30,* with the possible addition of a payor code for those client/transaction pairs which are not a priori the responsibility of a predetermined payor.

FILE 12.10.10 (TAPE FORMAT)

Field No.	Size	Data Item
1	7 (9)	Transaction Code
2	10 (X999999999)	Client Identification (Modified)
3	5 X	Provider Identification Code
4	X	Payor Identification Code
5	nX	Diagnosis Code

6	k (mX)	Treatment Code (s)
	8 (99999.99)	Treatment Cost
7	1X	PSRO Flag (0=No; 1=Yes)
8	8 (99999.99)	(Invoiced-Adjusted) Amount
9	1 X	HCDS Flag (1 = Overtreatment; 2 = Undertreatment; 3 = Client Complaint; 4 = Administrative Complaint)
10	8 (99/99/99)	Date of Service

Transaction Summary File

Records (or blocks)
may be sorted on any of the
fields depending on the
nature of the summary sought.

Now that we have defined the files and forms employed by the HCDS information system, we can move on to the last of the points of inquiry—the data processing manipulative logic.

THE FLOW PROCESSING LOGIC

There would really be no point in providing the reader with all the data processing logic diagrams or in attempting to detail all of the information handling functions. As must be clear from the illustration of the forms and documents—and of the overview of the data dimension provided earlier in this chapter—the majority of processing tasks are simple and straightforward. No new technological barriers are being broken, and there are no special machine configurations required to implement the HCDS design. Therefore, this volume will conclude with a very brief analysis of two of the central aspects of the data processing operations, the *referral processing* logic, and the information manipulation routines that are associated with *transaction handling.* The reason these two functional areas were chosen is this: they are requirements that are found in virtually every social service delivery system, and involve operations that are not likely to be overly familiar to the system analyst whose experience has been with industrial concerns or with the traditional governmental enterprises.

The data processing dimension of the referral process responds to the substantive logic reported in the last section of

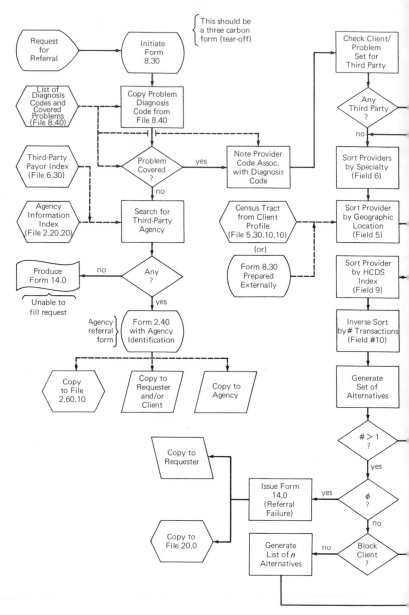

CHART 5.2 / Data processing operations related to the referral mechanism

Chapter 4, and indeed serves merely to implement that logic. Knowing something about the various forms and files that are generated and manipulated by the HCDS information processing subsystem, master logic Chart 5.2 should hold no surprises.

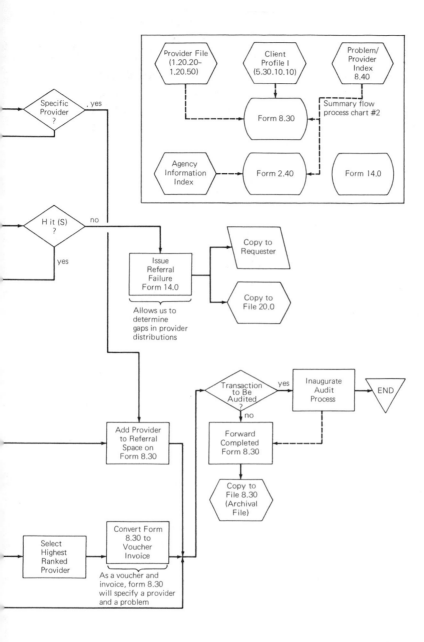

The data processing demands evoked by the referral process begin with a request for a referral by some client or other agent (e.g., a provider, a social worker). This causes us to initiate the development of a Form 8.30, the combination *referral notice*

and *voucher.* The list of diagnosis (condition) codes from File 8.40 is accessed, and the appropriate code inserted into the referral form. We then ask if this is a problem covered by HCDS. If not, then we search the agency information index (File 2.20.20) for some other agency which might take responsibility for the condition. If there is some other agency, we then generate a cross-referral (using Access Form 2.40) and alert the agency or at least provide the client with the necessary contact information.

If, however, HCDS does cover the problem, then we follow the usual processing trajectory. We first note the provider code associated with the condition, using File 8.40 to identify the specialty involved, if appropriate. We then check the client's profile for any third-party payor coverage for this condition. If there is such coverage, then the referral form will note who is responsible and also absolve HCDS from reimbursement. We then ask if the client has indicated a specific (preferred) provider. If the answer is yes, then the specific provider (subject to his being acceptable) is noted on the referral form. If there is no preferred provider, then the provider match *algorithm* is kicked in to isolate the most favorable provider, given the cost-effectiveness, geographic and distributive criteria. The search, or sort, is made first on the basis of specialization, next on geographic proximity to the client, then on the basis of the cost-effectiveness (HCDS audit) index. Finally, among those which remain at this point, an inverse sort is made on the basis of the number of previous transactions in which the providers have been involved. The result is the selection of a provider (or providers) who have the least number of historical referrals, but are similarly qualified on the other dimensions.

In some cases, no appropriate provider will be found, which results in the generation of a referral failure form. These, when collected over a period, suggest what medical services are most required but currently not provided; moreover, when geographic dimensions are employed, a variation on the failure provision can also tell us where the most urgent areas for additional providers are located. When there is success with a referral, we then generate the referral, and copy the holding file and the requester. For block patients, we simply convert the referral notice into a voucher and provide the client with the one-shot treatment he requires and with a specific (and exclusive) provider authorized to deliver the treatment. For patients covered under other modalities (especially in the fee-for-service and pre-paid category), a list of alternatives might be provided, assuming that more than one provider emerged from the selection algorithm. At any rate, the client is now released to seek treatment, with Form 8.30 as his authorization for block clients, or as a decision alternative (informational input) for other clients.

We now wait for the return on an *invoice* from a provider (Form 12.10.30), indicating that some service has actually been delivered. This sets off the normal *transaction handling* routines, as illustrated in logic Chart 5.3. As can be seen from this chart, the first step in the transaction handling logic is to test the client's status within the HCDS system. This is done by noting the code in field #2 of the Client Profile II (for the codes and their implications, see again field #22 of the format for File 5.30.-10.10). If the code is 4, this means that the invoice is invalid because the provider's patient has been rejected by HCDS; in this case, we would send the provider a Form 12.50, noting nonresponsibility for reimbursement (as would be done if a provider not approved by HCDS submits an invoice). If the code is 6, this indicates that the client is on contingency enrollment and that the reimbursement request will be held in File 12.80.10 until a disposition is made. In some cases, should a program implement the special strategy earlier mentioned, the provider would be paid forthwith, and the billing held in an account receivable should the client subsequently be excepted by HCDS. We then check to see that the problem for which treatment was given is indeed covered by HCDS; if not, then we would again forward a Form 12.50 (notice of nonresponsibility). These various tests in the transaction handling routines are here because we do not want to have to tie reimbursements to specific referrals. That is, many clients in the nonblock modality will be making their own arrangements; for them, the referral process is an optional service and not a necessity. Therefore, not every legitimate treatment or provider contact will be preceded by a referral.

Moving on, the rest of the processing steps neatly parallel the logic set out in Chart 5.1. The client and provider records are checked for copayor status, for third-party payors, etc., and a residual HCDS responsibility determined. Where, for example, a third-party payor is deemed responsible, a copy of the invoice is sent to it, and the cost for treatments zeroes out of the client's record. Finally, we make arrangements to enter any data that

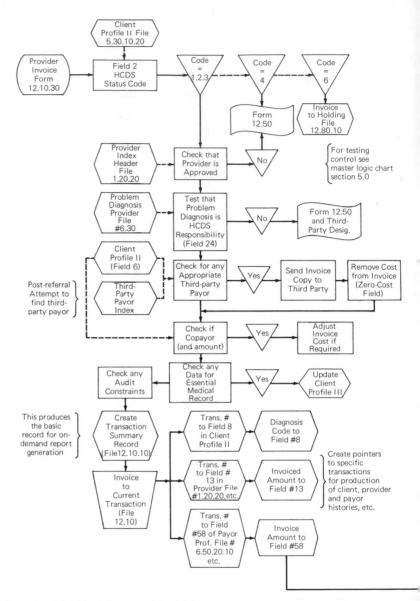

CHART 5.3 / Transaction handling routines

might be appropriate for the client's essential medical record (File 5.30.10.30), and note on his administrative record (in File 5.30.10.20) the condition treated, the provider and the cost of the treatment, and the transaction code which will subsequently allow re-creation of case histories on demand. As a final step, the

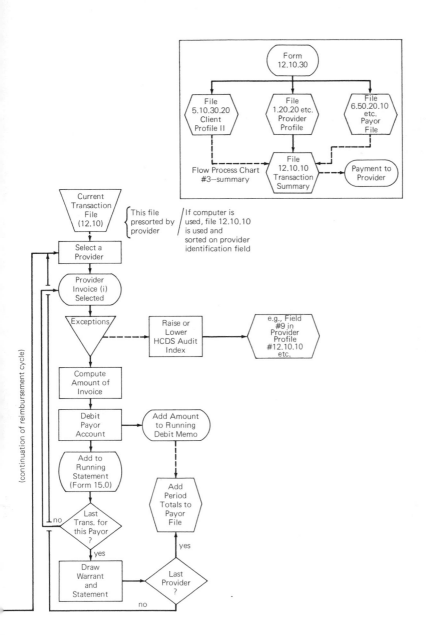

Form
12.10.30

File
5.10.30.20
Client
Profile II

File
1.20.20 etc.
Provider
Profile

File
6.50.20.10
etc.
Payor
File

Flow Process Chart
#3—summary

File
12.10.10
Transaction
Summary

Payment to
Provider

Current
Transaction
File
(12.10)

This file
presorted by
provider

If computer is
used, file 12.10.10
is used and
sorted on provider
identification field

Select a
Provider

Provider
Invoice (i)
Selected

Exceptions

Raise or
Lower
HCDS Audit
Index

e.g., Field
#9 in
Provider
Profile
#12.10.10
etc.

Compute
Amount of
Invoice

Debit
Payor
Account

Add Amount
to Running
Debit Memo

Add to
Running
Statement
(Form 15.0)

Add
Period
Totals to
Payor
File

Last
Trans. for
this Payor
?

no

yes

yes

Draw
Warrant
and
Statement

Last
Provider
?

no

(continuation of reimbursement cycle)

provider is sent his payment warrant, and the appropriate payor
account(s) debited. When this has been accomplished, the internal management logic described earlier would be inaugurated,
and the HCDS system has now run one full cycle—from referral
to reimbursement to fiscal implication.

At this point we can conclude this inquiry into social service management. Many of the arguments presented—and perhaps a majority of the techniques and instruments discussed—will perhaps wait another day for implementation. As was suggested, the current method of structuring the social service sector itself, and the problems with the educational preparation that many social service aspirants receive, must be revised and expanded, respectively, before we can legitimately anticipate the rationalization of social services. And this means that more literature must be made available from those involved in the social service sector, both as scholars and practitioners.

Considering the problems and the pain that inefficiencies and errors in the social service sector cause, neither the scholar nor the practicing social service manager can afford to be too tactful, too reticent. It is my belief that it is not pride that causes people to write books like this, but pride that keeps them from it. For while there is an element of arrogance about any author who tries to tell people what's wrong with some system, how it should be corrected, the reader must recognize that this very arrogance of ambition makes the author vulnerable to criticism and attacks which can sometimes be cutting and painful in their own right. It has been suggested that it is always safest to say nothing. I agree. But, as this volume testifies, I believe that safety and self-interest have little place in the calculus of the modern scientist, and no place at all in the calculus of those who feel that science should be of use to society.

I want to make one more point in closing. If I were really discouraged about the quality and character of those entrusted with our social service programs—if I were truly dispirited and believed every one of the criticisms I raised—then this book would not have been written. In short, the very fact of the criticisms is the most eloquent testimony I can offer to the legion of well-qualified, humble and effective social service functionaries. For them, criticisms cease to become criticisms, per se, and instead become contributions.

NOTES AND REFERENCES

[1] For a note on some of the structural implications of maximizing information leverage, see John W. Sutherland, *Administrative Decision Making: Extending the Bounds of Rationality* (New York: Van Nostrand Reinhold, 1977). For a technical treatise of great interest, see

also Borje Langefors, *Theoretical Analysis of Information Systems* (New York: Barnes and Noble, 1969). Finally, many points that system analysts might consider in regard to leverage are scattered throughout the fine book by Adrian McDonough, Information *Economics and Management Systems* (New York: McGraw Hill, 1963), especially Section II.

[2] This point emerges, most clearly, in the area of real-time and control engineering, cf., F. G. Shinskey, *Process Control Systems* (New York: McGraw Hill, 1967).

[3] The ideas are probably implicit in the work of most serious information system designers. For example, most systems designed for business applications presume a decision orientation. See, in this respect, chapters 5 and 6 of M. J. Alexander, *Information Systems Analysis: Theory and Applications* (Science Research Associates, 1974).

[4] Decision utility for information may be imputed to be the reduction in expected value of decision error that accompanies the generation of an increment of information. For more on this, see Chapter 5, John W. Sutherland, *Systems: Analysis, Administration and Architecture* (New York: Van Nostrand Reinhold, 1975).

[5] For an erudite and impressive analysis and exposition of normal governmental budgeting and accounting techniques, see Lynn & Freeman's *Fund Accounting: Theory and Practice* (Englewood Cliffs, N.J.: Prentice-Hall, 1974).

[6] The matter of opportunity costs crops up constantly (though perhaps often only referentially) in disputes about the economics of public or nonprofit enterprise. The reader might wish to turn to the index of North and Miller's book *The Economics of Public Issues* (New York: Harper and Row, 1971), and note how the concepts are employed in the several economic contexts they discuss so ably and with such interest.

[7] What I am suggesting here is that a priori (subjective) and a posteriori (empirical) probabilities may be cojoined, or more specifically that a posteriori probabilities may sometimes be used to modify (beforehand) probability distributions that are predominantly judgmental or subjective. This is expecially important in the context of Bayesean-driven dynamic decision making. In this regard see Martin J. Beckman, *Dynamic Programming of Economic Decisions* (New York: Springer-Verlag, 1968).

[8] Once any seasonal or periodic enrollment patterns are noted, certain standard time-series analysis instruments may be brought in to develop a projected demand function. See, for the technology, Chapter 14 of Croxton & Cowden's *Applied General Statistics* (Englewood Cliffs, N.J.: Prentice-Hall, 1955).

INDEX